048340

AF744219

Systems Engineering

Pages to Read

32
38
47
49

Systems Engineering

An Introduction

JOHN BOARDMAN

Prentice Hall
New York London Toronto Sydney Tokyo Singapore

First published 1990 by
Prentice Hall International (UK) Ltd
66 Wood Lane End, Hemel Hempstead
Hertfordshire HP2 4RG
A division of
Simon & Schuster International Group

Typeset in 11/12½ pt Times
by Best-set, Hong Kong

Printed and bound in Great Britain at the
University Press, Cambridge

British Library Cataloguing in Publication Data

Boardman, John *1946–*
Systems engineering.
1. Systems engineering
I. Title
620.0042

ISBN 0–13–504416–2
ISBN 0–13–504424–3 pbk

1 2 3 4 5 94 93 92 91 90

I would like to dedicate this book to my wife Renee,
and to our three children Richard, Sarah and Jonathan

Contents

Foreword

Engineers develop models of complex systems in order to understand their behaviour and to shape their design. This book is about modelling systems involving the behaviour of multiple, interrelated physical subsystems embedded within larger social and economic systems in which the physical systems must function. Engineering education and practice have traditionally focused on modelling the behaviour of physical systems using the tools of continuous mathematics. Until the mid 1960s, the solution of all but the most commonplace engineering problems required that engineers formulate and solve systems of differential equations. Since the late 1960s, the widespread availability of CPU cycles on ever more powerful and less expensive digital computers, coupled with rapid advances in approximate or 'numerical' mathematical solutions to systems of differential equations transformed the practice of all branches of the profession.

During the 1970s database management systems facilitated the organisation and input of massive amounts of engineering data to describe large, complex physical problems – everything from huge structural frames to tiny integrated circuits – and enabled standards to be developed for machine-to-machine communication of intermediate solution states between analysis and design programs. At the same time, advances in computer graphics augmented database management systems to enhance man-machine communication for engineering problem definition and for interpreting the output of engineering analysis and design programs. These advances in computer science and technology have extended the ability of engineers to perform analysis and design within a single discipline. They have created powerful 'islands of automation' within which specialists have improved their productivity and extended the classes of systems that they can analyse from the perspective of their own disciplines.

Yet engineering is inherently interdisciplinary. Space shuttles, global telecommunication systems, Channel Tunnels and large-scale integrated circuits all require input from many engineers in a variety of disciplines, and engineered products are frequently subsystems embedded in larger systems with whose components they interact in increasingly complex ways. The advances in computer algorithms, database management systems and graphics that have proven so useful to designers in optimising their small pieces of such systems have so far contributed only marginally to the integration of design perspectives and engineering subsystems. The consequences of this dichotomy are sputtering productivity gains in many engineering product sectors and increasingly brittle engineered systems. Witness the West Gate Bridge collapse in Australia, the Challenger space shuttle disaster in the United States, the Zeebrugge ferry sinking in Belgium and the catastrophic Chernobyl power plant explosion in the Soviet Union, all within the last few decades.

The discipline of 'Systems Analysis' bloomed in the post-war period fuelled by engineers who had invented radar, advanced weapons and other multifaceted engineered systems. Early work on systems analysis followed the style of engineering analysis within individual physical subsystems. Complex systems were described in terms of numerical objective functions and constraints, and computers were widely employed in the solution of the large systems of equations that were used to describe physical, social or economic systems. In spite of significant progress in operations research and systems analysis since the 1940s, the more subtle interactions among the elements of large complex systems frequently involve qualitative objectives and constraints and have proven to be less amenable to solution by such methods.

During the 1980s, a new breed of computer scientists developed a set of representation and reasoning techniques for modelling large complex systems qualitatively. Early work in knowledge-based systems or 'expert systems' relied heavily on superficial or associative heuristics to perform diagnostic tasks, first in medicine and subsequently in a variety of engineering applications. During the second half of the 1980s work by computer scientists and engineers, often working together in teams, has led to a different style of qualitative reasoning – now termed 'model-based reasoning' – in which components of engineering systems are modelled explicitly and are assembled into system models in which structure, function and behaviour are represented. Such systems can be exercised to simulate in both qualitative and quantitative ways the physical/social systems that they model, and can be used as the kernels of reasoning systems performing tasks ranging from design, through scheduling, to real-time monitoring and control, condition evaluation and maintenance planning.

Professor John Boardman's intellectual odyssey through the domains of electrical engineering, systems engineering and, most recently, artificial intelligence, have provided him with a unique perspective from which to synthesise these alternate approaches to modelling and shaping large complex engineered systems. *Systems Engineering* provides at once a thorough review and introduction to the tools of systems analysis, both quantitative and qualitative, and a rigorous and coherent approach for forming and shaping engineering models that can capture the intricacy and the subtlety of the man-machine engineered systems in the modern world of engineering.

If engineers continue to define their profession as the analysis and design of single physical systems from a one-dimensional intellectual perspective, they will rapidly become para-professionals, who merely provide data input to ever smarter analysis and design programs. This is already beginning to occur in some areas of engineering practice. The challenge posed by this book is for engineers to broaden their horizons and expand their intellectual toolkits so that they can assume a new professional role: engineering our society's complex, man-machine systems for agriculture, manufacturing, communication and recreation. Examples of these systems are already becoming large enough to interact in measurable ways with the increasingly fragile system called 'Planet Earth.' It is truly a heroic challenge.

Professor Raymond E. Levitt
Department of Civil Engineering
Stanford University
Stanford, California, USA

Preface

Systems engineering is an immensely practical subject that also offers a unique richness of approach to problem solving, since it includes a great variety of design tools and techniques coupled with a wealth of decision-making methods. The characteristics of the subject area with which systems engineering deals are the complexities of scale exhibited by the systems themselves, and the subtle relationships which exist between these and the problems with which they are intended to deal.

Since this book is by nature an introduction to the subject of systems engineering, its contents emphasize these three facets:

1. The subject's practicality.
2. The subject's richness of approach.
3. The intricacies of the problem-solving area.

The book's initial definition for the subject is as follows: 'Systems engineering is a rational approach to decision-making related to the solution of complex problems in engineering planning, design and operation.' It follows from this that the problem-solving area is particularly associated with large-scale systems, the type that serves large numbers of people. Examples include air traffic control (ATC) networks, coke manufacturing plants or, indeed, an organization such as a national rail network.

The book examines the meaning of this definition in the context of the claims of other subjects – such as operations research, management science and planning – to deal with the kinds of problem that systems of this scale and complexity commonly exhibit. A discussion of the international telephone system is presented to shed further light on the need for a more adequate resolution of complex systems problems.

Methodologies other than the traditional kind that serve hard systems engineering are examined. A robust definition of systems engineering is put forward. This takes advantage of the more recent and exciting ideas which emanate from the application of systems thinking to software engineering and artificial intelligence.

The book emphasizes the central importance of fundamental systems concepts such as *boundary*, *emergence* and *hierarchy*. Wherever possible these concepts are included in the description of the traditional design tools that are so essential to the systems engineer who is concerned with reliability, optimality and customer satisfaction.

A question which the reader will undoubtedly wish to put and have answered satisfactorily is 'What does *this* book have to say about systems engineering?' The book is not intended to be merely a commentary on other people's ideas although the author has of course benefited from a detailed consideration of these. The book makes its own statement of what systems engineering is, and what profit there is for the reader in receiving this message. It makes three important contributions in introducing the reader to the subject of systems engineering.

First, it provides a philosophical framework by which the reader can understand the need for systems engineering and its role within the wider spectrum of engineering. Secondly, it presents a collection of formal techniques, which, when grasped with competence and confidence, will enable the reader to make effective and practical progress with the exercise of applying systems engineering to his own area of interest. The book attempts to convey these techniques through clear examples and structured exercises, so as to impart the necessary degree of competence and confidence efficiently. The reader will gain confidence in these techniques through their regular use, and will thereby acquire new skills in the respective situations. Finally, the book shows how systems engineering relates to the emerging subjects of software engineering and artificial intelligence.

A cornerstone of this book's systems approach is the set of definitions which identifies three key concepts: system boundary, emergent feature and hierarchical interdependence. These concepts are drawn out and illustrated in contrast with other system ideas in Chapter 2.

An essential prerequisite to the understanding of many design components is a thorough grounding in probability theory. The book therefore starts with a debate on coping with uncertainty and handles this in a mathematical sense in Chapter 3. It continues its exposition of the systems engineer's principal tools with a discussion on reliability in Chapter 4. The importance of the prospect of systems failing is recognized here, for even if systems are small-scale and non-complex, failure is possible and must be a prime consideration for the systems engineer. By virtue of becoming large-scale and complex by nature, necessarily incorporating the human element somewhere within them, systems will inescapably be confronted by failure in some form or fashion. Failure can both be anticipated (by building in redundancy based upon reliability analysis) and expounded upon retrospectively; the latter requiring skills for effective post-mortem examination, particularly following some catastrophe.

The set of design components is completed by chapters on queueing theory (Chapter 5), linear programming (Chapter 6) and dynamic programming (Chapter 7). In order to give the budding systems engineer an insight into the management of systems, Chapter 7 includes a description of the use of networks as a modeling technique for the purpose of project management by critical path analysis (CPA). This further describes general network modeling techniques and heuristic methods which are forerunners to much of the effort being put into artificial intelligence.

Finally, in Chapter 8, other skills, necessary for the effective exposition of engineering expertise, are covered. These include accounts of recent developments in the area of intelligent knowledge-based systems, which are likely to give systems engineering even greater prominence in the future. A summary is included of projects in which the author has been personally involved which hopefully shows the efficacy of the techniques covered by the book.

A number of people, too many to mention individually, deserve acknowledgment in respect of the development of this book. Mary Pollard, Jill Armstrong and Lindsey Ashworth deserve thanks for helping with the manuscript. My friends and colleagues at Brighton Polytechnic, and particularly those of the Information Technology Research Institute, are all gratefully acknowledged. Nigel Brown deserves special thanks for his work on the index and proofreading. Stephen Lea, Chris Garrett, John Kellett and Alistair Morton also helped with the proofs.

Special mention is due the following people. Brian Swanick and Phil Mars were a powerful influence in my early days at the University of Liverpool. Denis Farquhar was a great help in getting me to understand how big real problems can be. Dave Sandoz gave me advice on which problems to go for, and Ray Wilde, Ian Richards and Geoff Francis, who were my first Ph.D. students, helped in solving these.

Geoffrey Hall and Fulton Lee encouraged me to enter the world of information technology, a move I did not regret. Tim Barber and Garry Marshall performed quite excellently as my first partners in this move. Sandy Robson has supported me throughout the long period of getting my ideas down on paper.

John Boardman

1 Philosophical foundations

1.1 PRELIMINARY OBSERVATIONS

Systems engineering is an immensely practical subject, and this fact at the very least must be put across for an introduction to the subject to be effective. Any opening remarks should also reflect the richness of the subject's approach to problem solving, for it includes a great variety of design techniques and decision-making methods. The problem-solving area dealt with by the subject must be clearly established since it is, of course, important to the reader. Its significant characteristics are the complexities of scale and the subtle relationships between the systems which are studied and the problems which they pose. In developing the content for an introduction to the subject, therefore, emphasis is placed on three facets:

1. The subject's practicality.
2. The subject's richness of approach.
3. The intricacies of the problem-solving area.

A definition of the subject with which many commentators have readily agreed is:

'Systems engineering is a rational approach to decision-making related to the solution of complex problems in engineering planning, design and operation.'

It follows from this that the problem-solving area is particularly associated with large-scale systems, the type that serves large numbers of people, such as an air traffic control (ATC) network, Figures 1.1, 1.2 and 1.3, or an organization such as British Rail, Figure 1.4.

These examples clearly represent a wide range of systems, and it is perfectly reasonable to question whether 'systems engineering' is a fitting description for an approach to these and similar problems, keeping in mind the broad meaning of engineering. It may be that the subject constitutes a valid and feasible approach, but then others such as management science or systems analysis can make similar claims. Engineers are

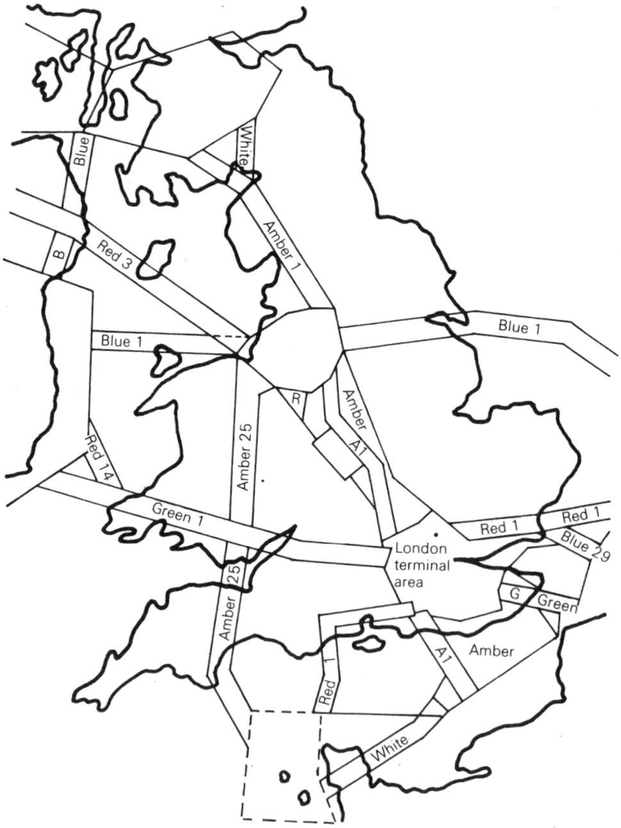

Figure 1.1 Motorways in the sky.

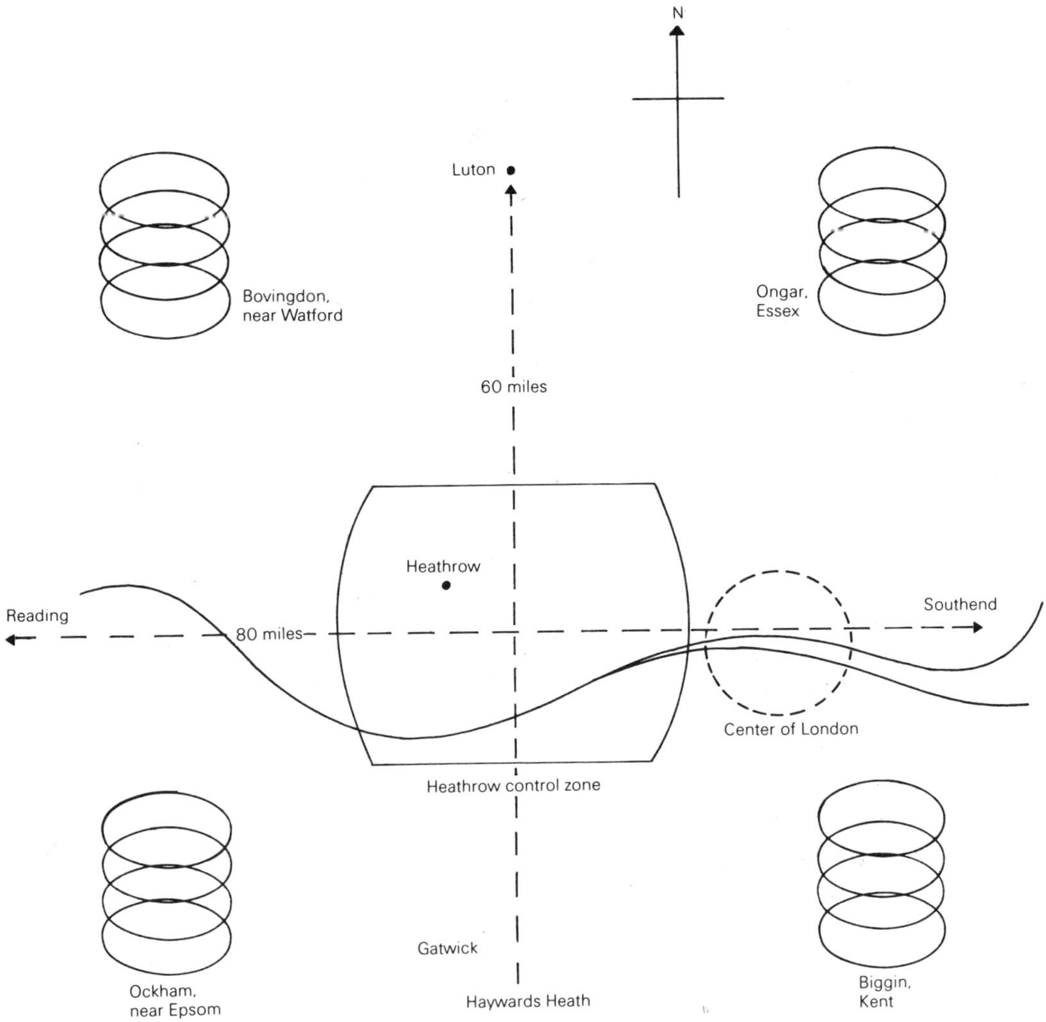

Figure 1.2 Using stacks to control air traffic for the London airports.

bound to wonder whether these other approaches, by virtue of their titles alone, are not more suitable.

The implication is, of course, that systems engineering may not be anything new but is just good engineering dressed up in new clothes. Such doubts, which are an irritable source of confusion and an undesirable influence on the engineering profession, should be dealt with swiftly and decisively. This calls for a re-examination of the philosophy of engineering itself.

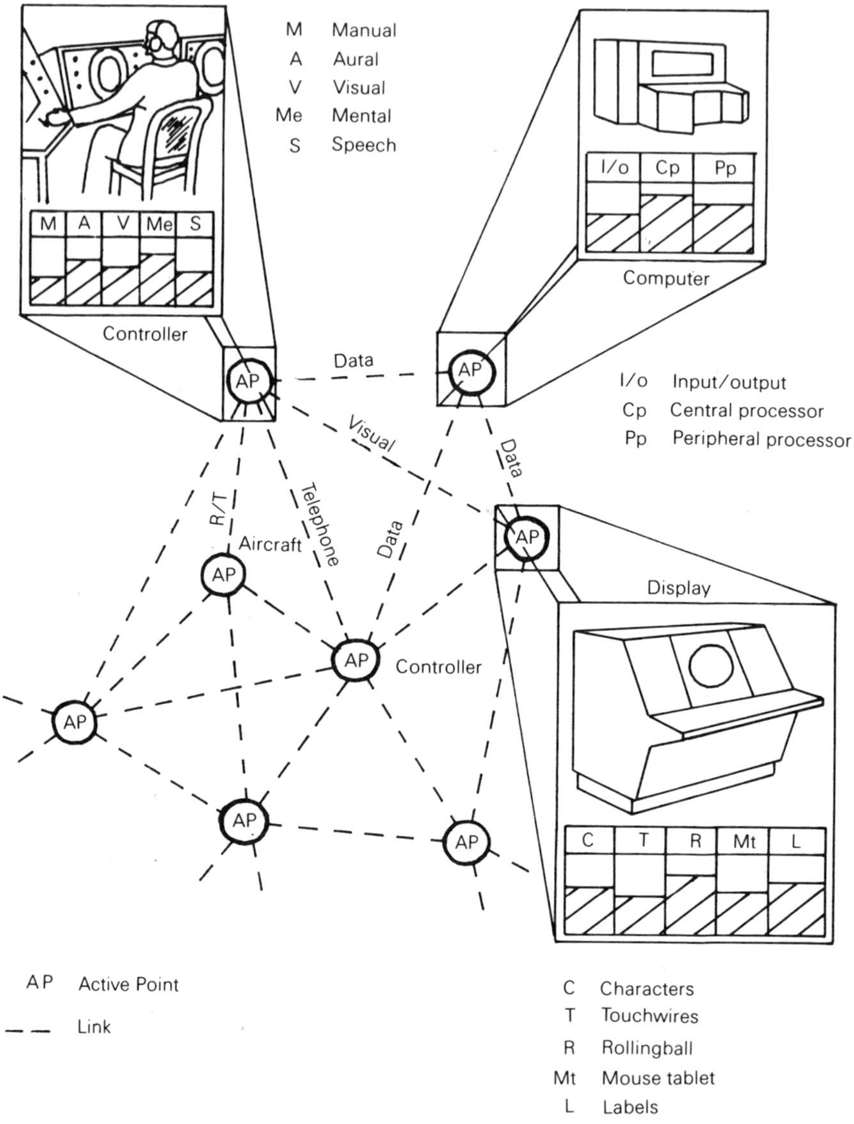

Figure 1.3 A model of an ATC network to simulate overall system performance.

1.2 PHILOSOPHICAL CONTRIBUTIONS TO ENGINEERING

In March 1961, a Committee of the School of Engineering at Massachusetts Institute of Technology, prepared a statement entitled 'Engineering and education' which included the following:

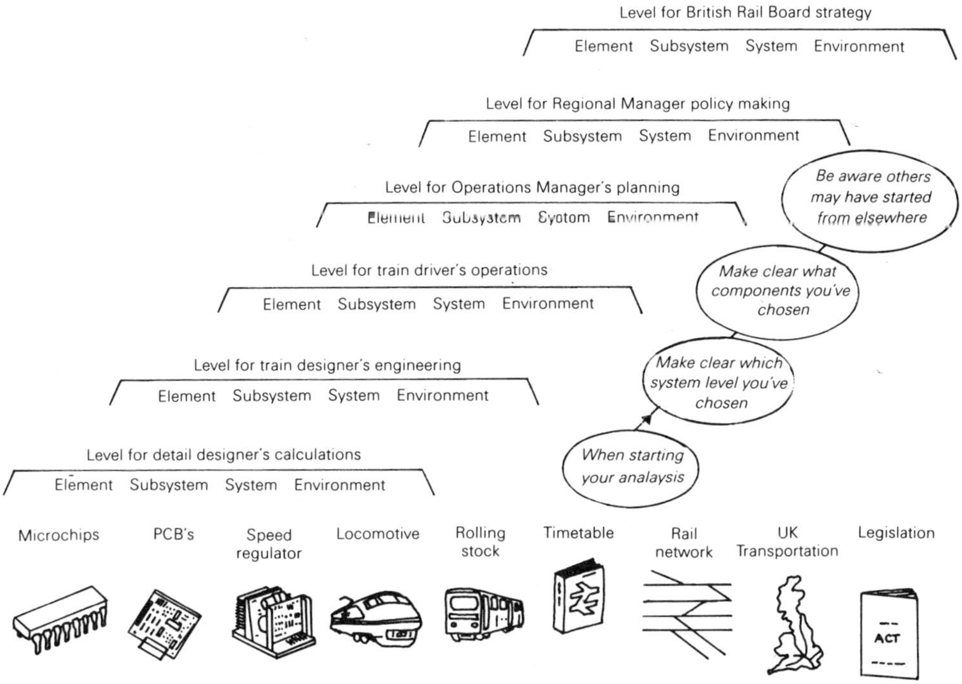

Figure 1.4 A systems view of British Rail. Taken from Paul Vernon (1987) 'A systemic approach to the design of systems', *Electronics and Power*, June 1987.

Engineering is a profession. Its members do creative work which results in things that people need or want. These things may be highways, submarines, interplanetary vehicles, antibiotics or television. Science, on the other hand, is a search for knowledge. The science of mathematics extends abstract knowledge. The science of physics extends organized knowledge of the physical world. In each of these, consideration can be limited to a carefully isolated aspect of reality. The engineer must deal with reality in all its aspects. He must not only be competent to use the most classical and the most modern parts of science, but he must also be able to devise and make a product which will be used by people. Moreover, he must assume professional responsibility insofar as the safety and well-being of people are affected by the thing he makes.

Dean Gordon S. Brown, in his delivery of the Medenhall Lecture at Harvey Mudd College, Claremont, California, on 7 May 1962 entitled 'The Engineer: Today's Pacesetter of Change', made the following remarks:

> Engineering is not merely analysis; engineering is not merely the possession of the capacity to get elegant solutions to non-existent engineering problems; engineering is practicing the art of the organized forcing of technological change – and this is something very different. Engineers operate at the interface between science and society. . . . When an engineer works at the frontier of his field his main function is to couple science – old science and new science – with his particular problem in order to build something and make it work. He must frequently make a commitment before he has ever demonstrated the feasibility of the objective. I support the argument that unless the prospective engineer is exposed during his undergraduate experience to the attitudes of engineering, he will never be properly motivated for his later training and professional career.

The implications of Dean Brown's comments that a philosophy of engineering needs to be infused into a young engineer's education in order for him to mature successfully into a professional practitioner are interesting, for it is one of the purposes of this book to suggest an orderly and expeditious integration of systems engineering into the undergraduate curricula for engineers and applied scientists. For the moment, however, we concentrate on an evaluation of the above remarks in relation to the position of engineering itself.

In his book *Foundations of Engineering* (1964), Philip Sporn comments on these remarks made by both the MIT Committee and Dean Brown:

> The engineer must often go beyond the limits of science, or question judgement based on alleged existing science. He must frequently assert his own over-riding judgement and stake his reputation to go into areas beyond that which has been fully explored scientifically and indeed, may even contravene that which may be claimed to have been demonstrated incontrovertibly by the science of the day.

Sporn paraphrases Justice Holmes's comments in *Portrait of a Philosopher* by saying 'Science and scientists have all been cocksure of things that were not so.' Sporn is particularly attracted to the succinct expression made by Eric Ashby in his *Technology and the Academics*: 'Technology is of the earth, earthy; it is susceptible to pressure from industry and government departments; its end an obligation to deliver the goods.' Some final extracts from *Foundations of Engineering*, in which Sporn articulates with profound accuracy and inspiring passion a philosophy for engineering, are now offered to underline his general statement about engineering and the engineer:

> In their simplest concept, technology and engineering are applied science. But engineering is really a great deal more than this. It embraces all human experience with science, tools, methods, systems, and social organisation that add leverage to man's effort and make possible much greater abundance than his unaided physical strength and skill alone could deliver. The engineer is the key figure in the material progress of the world. It is his engineering that makes a reality of the potential value of science by translating scientific knowledge into tools, resources, energy and labour to bring them into the service of man. To make contributions of this kind the engineer requires the imagination to visualise the needs of society and to appreciate what is possible as well as the technological and broad social understanding to bring his vision to reality.

1.3 THE RELATIONSHIP BETWEEN ENGINEERING AND DESIGN

The foregoing remarks should have served as a timely reminder of the force of engineering to deal with a broad range of society's problems. This may help us to accept systems engineering as a plausible approach towards the investigation and resolution of large-scale systems problems. In the general statement of philosophy for engineering, it is implicitly suggested that the design function is the hallmark of engineering as distinct, say, from science. Can we say then that, in essence at least, systems engineering is the design of large-scale systems? In looking at this argument it should be admitted that there are many engineers who would regard 'good engineering' and 'systems engineering' to be synonymous. Remarks that have been made to the author include: 'I still find it hard to convince myself that there is a difference between systems engineering and just good engineering' and 'Since engineers have always been concerned with developing systems, they always have been doing systems engineering as (synonymously) engineering design.... Design is the function which characterizes the engineer.'

G.M. Jenkins (1969) agrees that 'Design is the central activity in engineering and, therefore, that good engineering and good systems engineering go hand-in-hand.' However, he goes on to say that to take such a narrow view of systems engineering misses two very important points. First, in any situation in which an engineer is involved, there is *a systems problem* to be solved before *conventional* engineering can begin, and the detailed engineering design can turn out to be pointless and wasteful in resources unless the systems problem has first been properly understood and treated.

A second and more important point which distinguishes engineering and systems engineering is that 'the same systems thinking which can be applied to the design of hardware systems such as space rockets, plants or ships, can also be applied, for example, to parts of firms, or whole firms, or to local government.... Thus the systems approach to design can be applied to these much wider commercial, management and "software" systems just as well as to the "hardware systems" with which the engineer is more familiar.'

Jenkins's comments are very helpful because they underline the essential nature of systems themselves, i.e. their property of being embedded in larger systems, and they emphasize the particular qualities of a form of overall systems thinking, such as systems engineering, to deal with the complexities of scale and the interactions within systems problems.

His remarks, however, once again raise the question of how appropriate an approach entitled 'systems engineering', can be as opposed to, say, operations research, in dealing with those systems not traditionally associated with engineers. This question can only ever finally be dealt with by a hard examination of all forms of systems approach. In the climate of a rational debate, a great element of doubt and confusion can be removed by allowing a clear statement of each approach to be made in order that its opponents may state their opposition as cogently as possible.

To the singularly practical-minded engineer, the prospect of a lengthy philosophical discourse holds no great appeal, a fact borne out by the personal experiences of the author in attempting to lecture the subject to undergraduate engineers. To gain its acceptance with students, or to achieve its inclusion in the curriculum, requires constant argument that it is indeed concerned with 'real engineering problems'. Of course this argument, though arduous and often tiresome, can be beneficial. Unfortunately, when it is naively compared with 'full-blooded nuts and bolts', engineering, the subject can be made to look like an outsider or a strange bedfellow.

This often brings out the simple argument that 'philosophy is all very well in its place along with abstract mathematics and esoteric design language, but really we must get on with the actual business of engineering'. This argument is born of blinkered ignorance and ultimately is socially irresponsible. To actually 'get on' presupposes a sense of purpose and, moreover, a clear understanding of what one must get on with! There is abundant evidence to suggest that the hasty action-oriented engineer has got on too fast and with it got into a mess! On the other hand, hundreds of large industrial firms in transportation, power, communication and materials can all point to divisions or groups within their organization which operate under one of the following labels: 'operations research', 'systems science', 'systems engineering', 'systems analysis', 'systems planning' and so on. In essence the particular label chosen does not really matter. The avowed purpose of all of these groups is the same – to approach problems in the spirit of systems thinking.

1.4 SYSTEMS APPROACHES

In his book *The Systems Approach* (1968), C. West Churchman sets out a debate between several interested parties, each of whom advance their individual, distinctive approaches to the one central problem, namely the understanding of the systems in which we live. The main topic of conversation in this debate is science, and in terms of textual prominence the central figure is the management scientist. We shall now elucidate the main points which are made in *The Systems Approach* so that we may arrive at a clear understanding of what we mean by systems engineering.

It must seem to many that we live in an age of stupid decision-making. About all that the decision-maker seems to do is pick on one aspect of the situation and push that as hard as possible, arguing against his opponents on the basis that they are failing to sense the true situation. People tend not to want to explore problems in depth with their antagonists, or to take on the burden of believing that their antagonists are right. People are not like that.

In these circumstances, science has often served as an arbiter, for it possesses attributes which suggest that it can perform this role agreeably in the majority view. Basically, science is independently minded and its freedom from the influences of industry, politics, religion or whatever is fiercely defended by scientists. Its sense of purpose, characterized as a search for truth by observation and reasoning, gives it a wide appeal to those who seek absolute value-free objectivity. The idea therefore that the dispassionate and yet clear mind of the scientists can aid in decision-making is not new: Plato had it many years ago when he thought he could begin to design the underlying model of a city-state, as he did in his *Republic*.

More recently, the success of the scientific teams in the military in World War II was outstanding both in the United Kingdom and the United States. As a consequence, after this war there was a rush to apply the same kind of thinking, which was then called 'operations research', to various non-military problems and in particular to industry.

As the scientist's perspective widened, he began to think of his approach as 'the systems approach'. He saw that what he was chiefly interested in was characterizing the nature of the system in such a way that the decision-making could take place in a logical and coherent fashion and that none of the fallacies of narrow-minded thinking would occur. What eventually was to emerge from this period of scientific systems thinking was science's standard bearer in the approach to understanding the systems in which we live, namely the management scientist.

Let us compare a management scientist with someone very similarly named, though philosophically poles apart, namely the scientific manager, otherwise known as the efficiency expert. He is the person who looks for and nearly always finds inefficiencies in the different aspects of any organization. To him idleness, in workforce or equipment, large stocks of inventory or other such evidence of waste, immediately suggest inefficiencies which call for change in each particular case.

The efficiency expert is 'right'. Virtually without exception, the operations of any industrial firm, government agency, public or private institution are inefficient and it is always possible to increase efficiency by changes in the work force or in work methods. The management scientist's argument against 'efficiency' is that it is always conceived in relation to a small segment of the social organization.

If the system is considered as a whole, one inefficiency has to be balanced with the others, and this is the point that the management scientist believes the efficiency expert (or scientific manager) misses. To label any one operation as inefficient is to miss the central idea of system thinking. To the management scientist, no one operation is inefficient by itself; the total cost of the system operation has to be calculated for each policy and this, says the management scientist, can be done, not by isolating one aspect of the system with a predetermined notion of what is best for it, but only by means of a model of the whole system. It very often happens that the efficiency expert's changes will worsen total system costs and that the management scientist's changes, leading to a reduction in total system costs, may run entirely contrary to that suggested by the efficiency expert.

For the management scientist, the systems approach entails the construction of 'management information systems' that will record the relevant information for decision-making purposes, and specifically will tell the fullest story about the use of a system's resources, including lost opportunities. His commitment, however, to this area of activity leads him into greater conflict with the other interested parties who would press their alternative approaches, particularly the humanist and the behavioral scientist. The reader is referred to *The Systems Approach* for details.

The management scientist has his critics, and by summarizing these criticisms the prospects for perceiving what might be regarded as *the* systems approach are increased. Let us therefore produce a synopsis of our management scientist so far. His credibility is based on the assumption that a very logical, astute, wise man can always get on top of a situation and look at the foolishness of people who are incapable of seeing the *whole*. However, we could only credit our management scientist with this ability provided that, first, we were absolutely sure that his objectives were pure and in line with those of the total system and, secondly, that we were certain that he actually possessed the appropriate observational powers.

An assessment of the management scientist's abilities can be obtained from the criticisms made by his main antagonists. We recognize that our large-scale systems serve large numbers of people and, moreover, will need large organizations of people for its operation. It is essential therefore to see how the systems approach performs in relation to the human being. A discussion about such a relationship soon becomes a debate about values, namely, the values of those who wish to change human systems by means of management science, by some other intellectual method, or by other means (described by Churchman as the auto-planning approach).

The management scientist has values that are explicit, precise and rational. They are most closely allied with those of the practical-minded manager who lays his problems out in an explicit form. The more elusive, intangible parts of values, therefore, are to be ignored or else handled by personal judgment. The values of his opponent, the humanist, are the deeper individual human values arising from each person's emotions. The humanist maintains that his values, in which of course he believes with sincerity equal to that of the management scientist, cannot be determined by any kind of intellectual investigation.

The management scientist wonders if he can answer this particular point, that his application of science does not adequately model human values, by combining forces with his scientific colleague, the behavioral scientist, to give a coalition approach to systems and one, perhaps, that will satisfy everyone. The behavioral scientist is an individual who believes that, by observing how people behave in their environment, it will be possible to describe their minds, ambitions and frustrations, and eventually to see how these fit together. Once the empirical investigation has reached a successful end, then the nature of human society will be understood; consequently the basis for the design of human societies will be the hard-core data about human behavior, rather than the assumptions of models. One would think, therefore, that the two types of scientist complement each other.

The behavioral scientist provides a real empirical base and the management scientist provides the structure that can employ this base for making inferences about the changes in social systems. The fact is, however, that the two are not well integrated and this is probably caused in the main by the basic psychological differences between the types of intellectual entering the separate fields. It is the difference between rationalism and empiricism. Attempts have been made to reconcile the two but none has as yet proved successful, demonstrating further the deficiencies in the approach of science to the systems problem.

The popular idea of how to approach a system is to get somebody to manage it. The manager is supposed to be a person with wide experience in the system and with a perceptive, brilliant mind. Management is a matter of experience and sound judgment and its quality can be judged in terms of performance. This point of view totally precludes using scientific method to achieve the required system performance, for it is exclusively dependent on the practical approach of experience coupled with intuition, leadership and brilliance.

Another type of anti-planner is the skeptic who firmly believes that not even the minor aspects of a system can be understood, let alone the whole. The skeptic believes that all that is said about systems is largely a myth invented in order to carry on various conversations and entertainments. Since the nature of the real world is a mystery, we deceive ourselves, maintains the skeptic, when we think we are improving anything. The skeptic is the ultimate pessimist and arrogant with it. The one thing he rarely does is defend his own skepticism. To the serious-minded, the skeptic's form of relativism serves little purpose and is socially irresponsible. This attitude of course in no way deters the skeptic. His approach to systems is that there is no sound approach, and that is that.

A more serious type of anti-planner is the determinist. He believes that major human decisions are not in the hands of human decision-makers but in uncontrolled sociological forces. The determinists often regard the whole attempt to study systems by means of science as a natural product of a highly militarized and industrialized society, thereby subsuming the management scientist in his own theory. The management scientist, on the other hand, will regard the determinist as someone who has arrived at his position because of frustration.

Churchman considers two further types of approach, essentially distinct from the intellectual systems approach, and these are the religious and the analysis-of-self. These are not discussed here for it is felt that by now the reader will have concluded that no single approach is *the* correct approach to the understanding of the systems in which we live.

We began this philosophical discourse with the straight comparison of the management scientist and the scientific manager. The former thought that the latter was being deceived. When the scientific manager sees idleness and slack in the system, he believes that he is looking at reality, while the management scientist maintains that he is seeing an illusion. But then the management scientist, when he becomes very serious about his own models, in which 'all' of the objectives are represented and a 'proper' compromise is created, is also deceived. In the straight-faced seriousness of his approach, he forgets many things: basic human values and his own inability to understand all aspects of the system, especially its politics.

Churchman believes that the attainment of a correct systems approach is itself approached by the repeated activities of perception and deception calling for a continuing reviewing of the world, of the whole system and of its components. 'It is nonsense', he says, 'to expect that any human being has yet been able to attain such insight into the problems of society that he can really identify the central problems and how they should be solved. The systems in which we live are far too complicated as yet for our intellectual powers and technology to understand.' Churchman feels this deeply and though himself obviously a supporter of the scientific view of the systems approach, manages to convey its complete inefficacy by saying 'There is something fiercely ugly at times about proposal writing, testing, and implementation of the programs of large-scale systems. It's much as though the whole life of the system had been depleted in an attempt to put it into a rational mode.'

Churchman's analysis described in *The Systems Approach* has many useful lessons for the would-be designer of an alternative approach to systems thinking. As we move on to make our own description of systems engineering complete, it is essential that we learn from these lessons. In conclusion of this philosophical discourse, therefore, we set down the principal points to emerge from this analysis:

1. The systems approach is the most critical problem we face today, the understanding of the systems in which we live.
2. The ultimate meaning of the systems approach lies in the creation of a theory of deception (Churchman, 1968) and in a fuller understanding of the ways in which the human being can be deceived about his world, and in an interaction the different viewpoints.
3. The essence of the systems approach is confusion as well as enlightenment; the two are inseparable aspects of human living. We have to maintain the contradiction or else we allow ourselves to be overwhelmed by the consistent (Churchman, 1968).
4. The systems approach begins when you first see the world through the eyes of another.

5. The systems approach goes on to discover that every world view is terribly restricted.
6. There are no experts in the systems approach. The problem of the systems approach is to learn what everybody knows.
7. The systems approach really consists of a continuing debate between various attitudes of mind with respect to society.

1.5 A SYSTEMS PHILOSOPHY FOR ENGINEERING

We are reaching the conclusion of these philosophical foundations. It is fitting to ask 'what have we learned so far?' and 'how can we embody this learning in a form that will equip us for succeeding stages?' These later stages will in general be concerned with a detailed examination of the contents of the subject, i.e. design techniques, and of the process of the subject, i.e. the action of its contents in formal application to valid problem-solving studies. It should be clear that what we mean by systems engineering, in principle, is as follows:

1. Some form of systems approach that has benefited from a careful consideration of the advantages and limitations of all other forms of systems approach.
2. Something more than what is meant or intended by 'engineering' even allowing for the fact that one form of systems approach is the engineering approach. (A note of explanation is necessary here. It is seemingly impossible to escape from the notions that, in the minds of a great many engineers, first, philosophy is too strongly subordinated for practice's sake and, secondly, the preoccupation with hardware systems seriously prejudices the effective consideration of software design and human element. These two mental characteristics impose a grave limitation on the efficacy of engineering as a systems approach. Nevertheless, our interpretation of systems engineering remains firmly practically oriented and faithful to the tradition of engineering.)
3. Something intrinsically design-inspired yet incorporating a sufficient degree of pragmatism so as to have a major influence on design practice.
4. Something that acknowledges a relation to the monumental and which therefore possesses a deep sense of the beauties and riches of the large-scale systems in our society, not only with regard to their spatial appearance but also in their abstract form.

The question we now put is, how can we encapsulate these thoughts of principle into a single statement which keeps faith with the original definition and yet declares the formative appreciation we have received through these basic observations and reasoning? We answer this question by making the following statement: 'systems engineering is a practical approach to the engineering of large-scale systems'. We are conscious that the particular structure of this statement has its meaning in relation to the emotive phrases 'practical approach', 'engineering' and 'large-scale systems' set out in the context of the foregoing discussion. We are saying the following:

1. The problem-solving area for systems engineering is now perfectly clear – it is the investigation and resolution of complex problems arising from the conception, articulation and implementation of large-scale systems.
2. We emphasize our faith in engineering as one approach in itself to the central problems of society, but we qualify this by recognizing the lessons to be learned from an examination of all such approaches.
3. Emphasis is to be placed on the need for an *approach* to the understanding of the systems in which we live and in so doing underline the power of modern science to this end, but that, in the final analysis, the real character of this systems engineering approach should be practical.
4. Finally, practical means: responsible; balanced; borrowing from the intellectual, yet observing the limitations of the rational; inclusive of the disciplines of engineering design but tempered with a common-sense attitude, respectful of human values and human feelings.

On this account, systems engineering is definitely not then another synonym for good engineering; rather it is a unique expression of engineering whose description is completed in a singular mission – the balanced investigation and resolution of complex problems arising out of large-scale systems.

1.6 THE MESSAGE OF THE BOOK

A question which the reader will undoubtedly wish to put and have answered satisfactorily is 'what does *this* book have to say about systems engineering and how does it propose to go about making that statement?' The book is not intended to be merely a commentary on other people's ideas although the author has of course benefited from a detailed consideration of these. The book makes its own statement of what systems engineering is, and what profit there is for the reader in receiving this message. The book makes three important contributions in introducing the reader to the subject of systems engineering.

First, it provides an original philosophical framework by which the reader can understand the need for systems engineering and its role within the wider spectrum of engineering. Secondly, it presents a collection of useful, useable techniques, which when grasped with competence and confidence will enable the reader to make effective practical progress with the exercise of applying systems engineering to his own area of interest. The book attempts to convey these techniques through clear examples and structured exercises, so as to impart the necessary degree of competence and confidence efficiently. It is written so as to enable the reader to gain confidence in the use of these techniques and acquire the skills to use them effectively in the respective situations. Finally, the book shows how systems engineering relates to the emerging subjects of software engineering and artificial intelligence.

It does this by means of a commentary on the central issues, examples which illustrate tools and techniques, and a brief case study in which the application of the tools and techniques, from both systems engineering and artificial intelligence, are considered side-by-side in the context of project management, paying particular reference to plan generation and project expedition.

The author believes systems engineering to be an organized collection of several interacting components for the design of systems, coupled with an important set of definitions which provide a framework of control structures for the use of the design components. The whole purpose of these design components and definitional controls is to tackle real-world problems known to be complex by nature and large-scale in size with a view towards identifying systemic solutions to these real-world problems. Notice that if the real-world problem is extremely ill-defined a systemic solution may be better known as problem definition. Other types of systemic solutions may be better known as repairs (to an existing system) or inventions. The philosophical framework which the book describes assembles these concepts and issues so that the reader can appreciate these definitions and design components. A wide range of real-world problems are included ranging from the softly defined problematical situation through to the sharply focused operational requirement. These problems reinforce the philosophical stance of the book.

A cornerstone of this book's systems approach is the set of definitions which identifies three key concepts: system boundary, emergent feature and hierarchical interdependence. These concepts are drawn out and illustrated in contrast with other system ideas in Chapter 2.

An essential prerequisite to the understanding of many design components is a thorough grounding in probability theory for which reason the book commences with a debate on coping with uncertainty and handles this in a mathematical sense in Chapter 3.

The book continues its exposition of the systems engineer's principal tools with a discussion on reliability in Chapter 4. The importance of the prospect of systems failing is recognized here, for even if systems are small-scale and non-complex, failure is possible and must be a prime consideration for the systems engineer. By virtue of becoming large-scale and complex by nature, necessarily incorporating the human element somewhere within them, systems will inescapably be confronted by failure in some form or fashion. Failure can both be anticipated (by building in redundancy based upon reliability analysis) and expounded upon retrospectively, the latter requiring skills for effective post-mortem examination, particularly following some catastrophe.

The set of design components is completed by chapters on queueing theory (5), linear programming (6) and dynamic programming (7). In order to give the budding systems engineer an insight into the management of systems, Chapter 7 includes a description of the use of networks as a modeling technique for the purpose of project management by critical path analysis (CPA). This further describes general network modeling techniques and heuristic methods which are forerunners to much of the effort being put into artificial intelligence.

Finally, in Chapter 8, other skills, necessary for the effective exposition of engineering expertise, are covered. These include accounts of recent developments, in the area of intelligent knowledge-based systems, which are likely to give systems engineering even greater prominence in the future. A summary is included of projects in which the author has been personally involved which hopefully shows the efficacy of the techniques covered by the book.

1.7 EXERCISES

1. Winston Churchill has been quoted as saying 'Engineers should be kept on tap, not on top', presumably referring to his opinion of their unsuitability to lead in any strategic sense in business, military or government. Discuss this remark and comment on what sound arguments a systems engineer might summon in order to defeat it.

2. Why does philosophy matter to an engineer? How should an engineer learn methodology? Describe the sort of education that would best prepare a person for becoming a systems engineer.

3. What must the engineering profession in the United Kingdom do in order to elevate the standing of the engineer to that of the doctor or lawyer? Is there a special role for the systems engineer in this exercise, and does the systems engineer deserve a prominent place in the hierarchy of the profession such as that of the consultant or barrister?

4. Describe the design of a pair of overalls for use by an electrician. How does this design process differ from that which would create a system for the mass production manufacture of overalls?

5. How would a traditional engineer, with no previous exposure to systems engineering in the formal sense, go about a typical systems problem such as the optimization of an airport? What mistakes would be made and what possible new insights could be gained to the benefit of the body of knowledge of systems engineering as a result of the traditionalist not having been 'blinkered' by the subject in the first place? Adopt a systems problem of your own choosing in order to develop a full description of the response that is sought.

1.8 FURTHER READING

Chase, W.P. (1974) *Management of System Engineering*, Wiley.
Checkland, P.B. (1981) *Systems Thinking, Systems Practice*, Wiley.
Churchman, C.W. (1968) *The Systems Approach*, Delta.
Flagle, C.D., Huggins, W.H. and Roy, R.H. (1960) *Operations Research and Systems Engineering*, Johns Hopkins.
Flood, R.L. and Carson, E.R. (1988) *Dealing with Complexity, an Introduction to the Theory and Application of Systems Science*, London: Plenum.
Jenkins, G.M. (1969) 'The systems approach', *Journal of Systems Engineering*, **1**(1), pp. 1–21.
Jenkins, G.M. and Youle, P.V. (1971) *Systems Engineering: a Unifying Approach in Industry and Society*, C.A. Watts.
Sporn, P. (1964) *Foundations of Engineering*, Pergamon.
Vickers, G. (1983) *Human Systems are Different*, Harper.
Yeates, D. (1986) *Systems Project Management*, London: Pitman.

2 Systems: cases and concepts

2.1 GENERIC PROBLEMS: GROWTH AND COMPLEXITY

Mankind is confronted by the two great unknowns: the origin of life and an indeterminate future. Hemmed in, as it were, on both sides, most people are prepared to accept that a basic fact of life is its uncertainty. What is certain, however, is that mankind has a history marked out by one great characteristic – survival. Dr Jacob Bronowski (1976) records this with inimitable charm and ingenuity as a cultural evolution whose brilliant sequences of peaks he labels *The Ascent of Man*. As he puts it, 'In body and in mind, man is an explorer of nature, the ubiquitous animal, who does not find but has made his home in every continent.... His imagination, his reason, his emotional subtlety and toughness, make it possible for him not to accept the environment but to change it'.

The survival instinct in man is indeed strong, for though by nature he may not be fitted to any specific environment, his survival kit equips him to cope with many environments. Surviving though is not a matter of plain-sailing. Rather, man's central problems can be viewed, simplistically at least, in terms of the struggle for survival. Broadly speaking, these relate to possessions, health, education, human freedom, war and the development of new resources.

This book is not intended as a comment on man's ability to survive nor is it a study of man himself. Primarily the book is concerned with a clarification of the peculiar complexity of these problems and of the massively complex systems which have been developed in response to them. Moreover the book goes to some length to present one particular approach, that of systems engineering, that may be used to establish those systems. As responses to the problems, these systems forestall the disastrous consequences that would result from failing to understand these problems correctly.

Man's progress through history has increased awareness in both breadth and depth of the existence of these problems but has further shown that this existence is not considerably less old than man. This companionship of man and his problems may indicate a static picture but it would be fallacious to see it as such, for change and the rate of change have come to exert significant influences in this situation. The history of man's survival is littered with changes; cultural, scientific, social, political and so on, and at every stage these changes have produced the legacies for successive generations. Man is no freer today of the consequences of his forebears' actions than they were of those of their own antecedents. In fact he is constrained to a greater degree because change and growth are considerably more powerful forces now.

The advances in technology and in man's knowledge of his environment become more noticeable with successive generations. Subjectively, this growing complexity would not have been noticed in the lifetime of a single individual 200 years ago. Today, the comments of an older generation indicate amazement at the change in pace and intricacy of life since their childhood. Objectively, this changing rate may be roughly measured by various statistics. First, there are more of us and the interaction between ourselves has increased considerably, and continues to do so. In addition, we move with greater speed, between cities and continents and, more recently, into space. The growth in all of these factors is indicated in Figure 2.1. Moreoever our daily lives are filled with an increasing number of functions performed with greater variety and sophistication.

The central problems of society today and this growth in human activity, endeavor and knowledge are inextricably linked. This linkage, called feedback by engineers, can be very destabilizing, particularly if the interdependency and the effects of alternative actions on it are not clearly understood. Furthermore, time is of the essence. Our dependence on the development of technology and of elaborate systems of human organization is much stronger today than it has ever been, and our commitment to technology and its products now almost total. It is true that only a small fraction of the world's population suffer from any consequences of

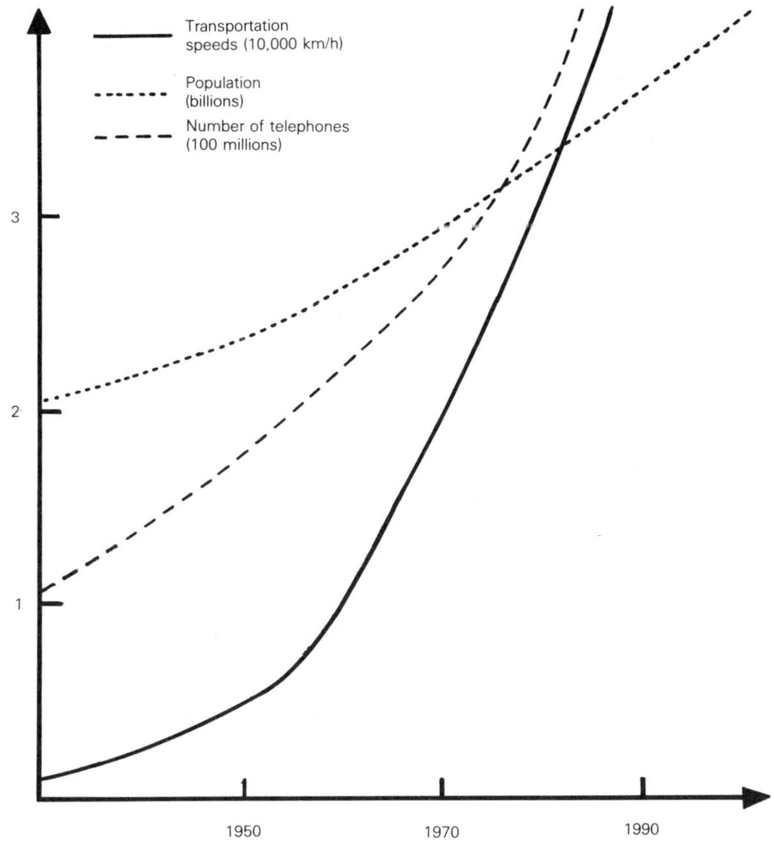

Figure 2.1 Growth patterns in human endeavor.

sophisticated life (noise, busy roads, etc.). In the developed nations, poverty, disease, poor education and restrictions on human freedoms are, to a great extent, no longer the dominant problems, at least relative to the conditions of, say, 200 years ago, but in large parts of the world, these are still prevalent evils.

However, even at the risk of exchanging these for congested roads, unabated violence in cities, and the frequent examples of human catastrophes resulting from failures in industrial or commercial operations, the desire among the leaders of these developing nations is towards a greater reliance on the products of modern technology and their attendant systems of organization. A simple yet effective demonstration of this widespread drive among the emergent nation states is the urgency with which they acquire a national airline. To fly one's own flag evidently achieves much for a nation from its birth. For example it encourages links with the rest of the world, thus stimulating the nation's economy; it affords the means of carrying its leaders to the UN for essential politicking; it shows the nation's flag in the major international airports of the world. However, of major significance, it is purchased, very often, in preference to the solutions for more threatening dangers such as widespread starvation or disease. The demand, therefore, for large-scale systems spreads throughout the world, which further underlines the importance that systems engineering should achieve success in coping with this demand.

2.2 SYSTEM TYPES: SOME ORIGINS

At this point it will be helpful to generate some ideas as to the variety and scale of the large-scale systems (both physical and abstract) which have been engineered in response to man's basic needs and requirements. In this way we can emphasize the interdependency of systems. To do this, we first add to man's basic need for survival the notion that by his efforts he can attain a more comfortable and satisfying life. Taken in combination, these two will extend our thinking to a simple classification of four system types which may be used to develop familiar examples for modern large-scale systems, Figure 2.2.

A first consideration is *safety* and there are two main objects in regard to which this must be established. First man seeks shelter from the elements of nature. He needs therefore to develop systems to support his health and welfare. Man needs to clothe, feed and protect himself and to maintain or repair his health. A second object is the class of artefacts which man devises. Safety must be a prime consideration in man's designs in order to minimize the risk of harm, injury or even death to those that may in any way be affected by those designs. A lack of regard for safety, therefore, would lead eventually to ill-health, catastrophe and, ultimately, the antithesis of survival, extinction. Perhaps not surprisingly therefore, planning and the means to carry out systematic planning are fundamental to the repertoire for achieving shelter.

A second consideration is movement. Perhaps it is man's basic desire for exploration that leads him to move about freely in his environment. In the modern world the stimulus for movement comes mainly from the dual activities of business and pleasure. The hunting instinct in man takes him on various travels to win the bread for the day. In contemporary parlance this is business, trade or commerce. Man's objective in seeking

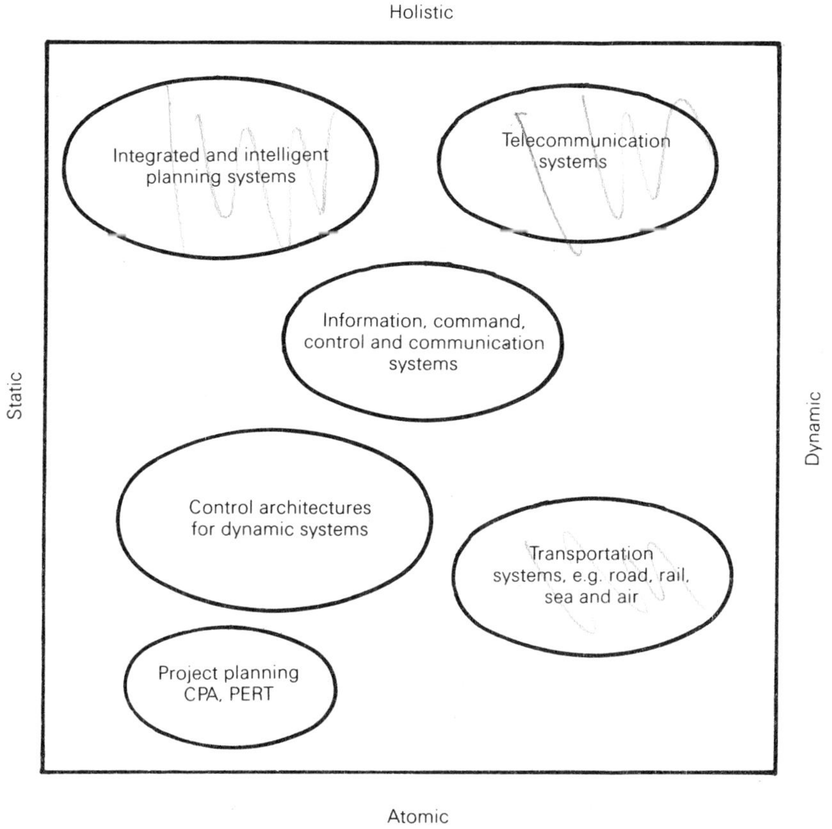

Figure 2.2

repose and refreshment from these hunts is today exemplified in pleasure travel, curiously enough itself a big business in today's world. The leisure industry, greatly stimulated by the world's airlines, is second only to oil in terms of financial turnover and exceeds fundings for armaments on the global scale. Transportation then gives us the second of our classifications with which to generate examples of large-scale systems.

A third type of system may be derived from a natural complement to transportation and that is communication. The history of man has seen quite startling changes in this facet of his adaptation to the environment. A few perfunctory thoughts tell the story in terms of the development of language, writing, printing, post, telegraphy, telephony, radio and television, bringing us finally to the sophisticated telemetry and telecommunication systems in the world today involving, among others, man-made earth-orbiting satellites, digital processing, optical fibers and lasers.

Our fourth and final category we call security and this is distinguished very clearly from our first category, that of safety. It is in the nature of man to fail. His basic failing is in his regard for fellow man. In our sense of good conduct, we all share, or should share, an awareness of injustice that is often drawn out in the relationships between members of the human race. Furthermore we all know of, if we have not personally experienced, the failure that so inevitably appears in the designs that man produces. Since man has been quite unable, by his own efforts, to find a cure for this failure phenomenon, his attitude to it has been as follows. First, he accepts that it will continue and that it is certain to pervade all that is done or made. Secondly, he prepares for the consequences of failure in two ways:

1. By anticipating its occurrence and thereafter providing some form of redundancy and damage limitation in his design approach.
2. By establishing protocols through which systematic examination of failure will expose the true cause of that failure and hence develop lessons for future improved designs.

Finally, he strives towards higher standards of security that will either reduce the likelihood of failure, or, in the event of failure, will at least minimize the effects of the catastrophe which results.

We shall now take a closer look at just one example in order to get hold of two main points. First, we should aim to discover some basic system concepts that are so much in evidence in large-scale systems and which a designer must firmly grasp in order to be sensible and practical in his designs. Secondly, we wish to appreciate the scale and complexity of the system in terms of its technology and organization, a combination of the scientific and the social. If we can do this we shall be enriched in our understanding of the central problems of society.

2.3 A COMPLEX SYSTEM: THE INTERNATIONAL TELEPHONE SYSTEM

The international telephone system (ITS) is a superb example of a truly large-scale system. It has over 300 million subscribers and covers more than 200 countries of the world. To describe it in the system vernacular would be to say that it is dynamic, adaptive, rapidly changing and yet stable and quite reliable. The system must be so organized as to be continuously available and on an economic basis, i.e. financially self-supporting. The main purpose of the system is to provide a transmission pathway between any pair of subscribers when they wish to communicate with one another.

As an example, let us take the case of a person calling a professor at the Massachusetts Institute of Technology from Brighton, Sussex, UK.

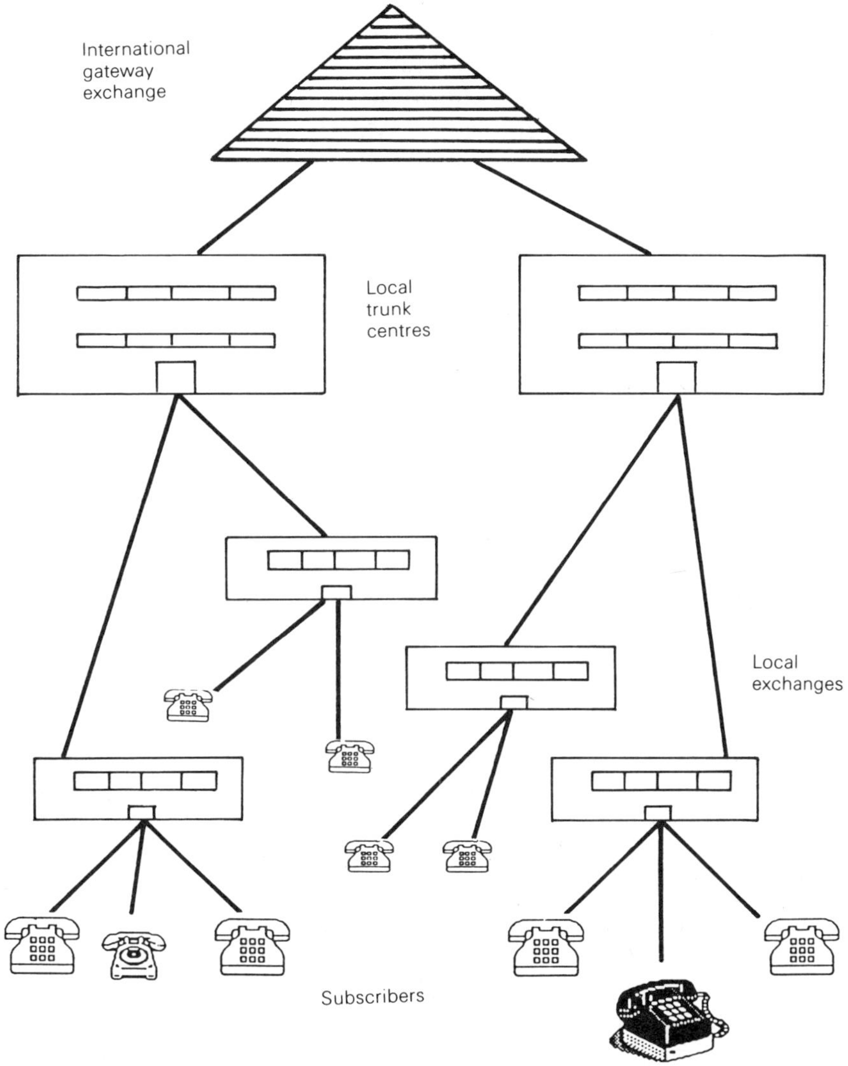

Figure 2.3

The professor's local number is 253 1992. The call is clearly not local to Brighton, indeed it is an international call. In fact the digits dialed are 0101 617 253 1992. The first three digits, 010, set up, quite automatically, a transmission path from the caller, via a local exchange and a local trunk center, Figure 2.3, to the international gateway in London. The next digit, 1, selects a path, one of several, between London and New York, in the United States. This link could be either a submarine cable or a satellite link and the selection is quite random. The remaining ten digits are then transmitted in sound-coded form along this path and the remaining part of the link completed in the United States.

The digits 617 represent the area code in the United States for the terminating caller in Cambridge, Massachusetts. Now the London to New York link is quite exceptional for the average international call since there is solid interconnection between these centers, or their international gateway exchanges. These gateways exist only in places where the demand for international calls is high. Let us take, for another example, the case of a call from Oslo, Norway, to Katmandu in Nepal. The relative unlikeliness of this type of call makes it uneconomic to provide a solid interconnection between these centers and it is very likely that this call will be made via the transit switching network that is rapidly being established as part of the ITS. This is a hierarchical arrangement of primary, secondary and tertiary centers, modeled very much along the lines of the British telephone system of local centers, local exchanges and the main group switching centers, Figure 2.4. There are seven primary centers at London, New York, Moscow, Bombay, Singapore, Tokyo and Sydney. These are fully interconnected, i.e. there are 21 solid links. There are many more secondary centers and examples of these are Paris, Nairobi and San Francisco. Oslo and Katmandu are examples of the tertiary centers.

There are usually several different ways in which a transmission path can be established on the transit switching network, although usually it is on the basis of the smallest number of links giving maximum transmission quality and minimum cost. In our example the pathway would be established via London, Bombay (this link being of high quality and high capacity) and then by high-frequency radio link to Katmandu.

An interesting observation from the systems point of view is that to any individual the value of the system increases with the size of the system, since it becomes very much more attractive when one is part of a larger, fully interconnected system. The effect of this is, of course, to create 'snowballing' situations where the system needs to grow in order to cope with demand and this, of course, leads to greater demand.

The growth of the telephone industry is, like many others, exponential,

Figure 2.4

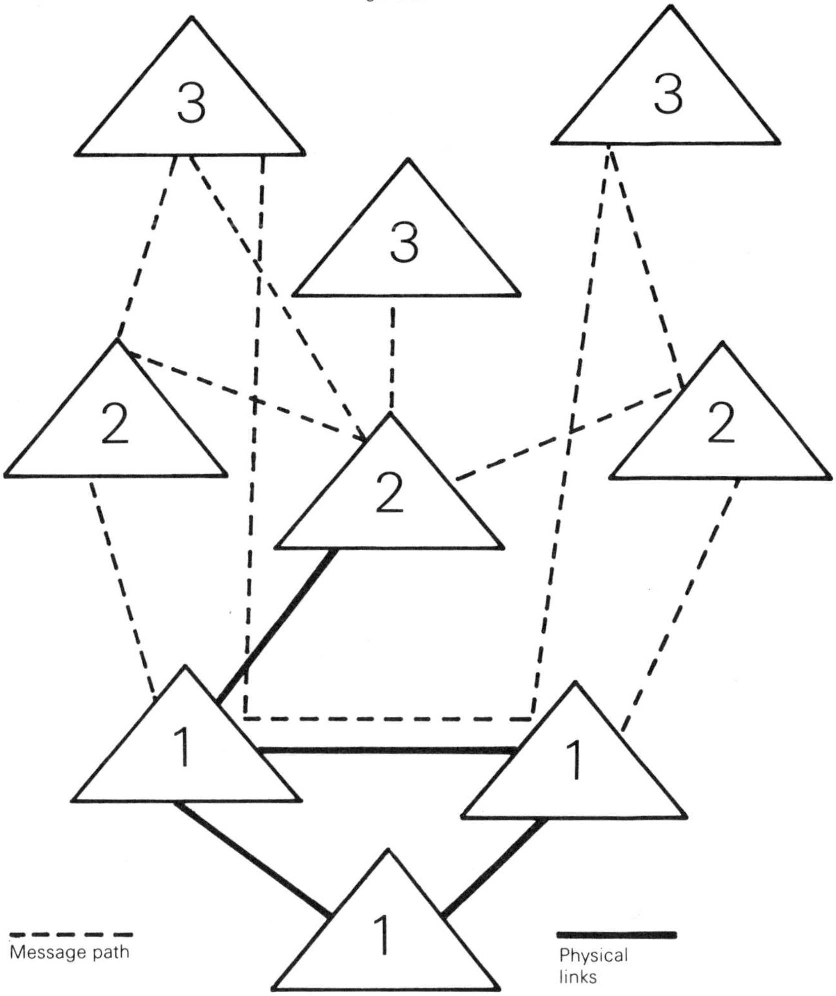

Figure 2.4

where over a period of just a few years the system has been faced with, and still faces, a sixfold increase in its own size. This very clearly poses a major planning problem and it is interesting to discover how the organization in charge of the network faces up to this problem. The ITS, organizationally, can be represented by the scheme shown in Figure 2.5. It is effectively made up of the international sections of the national organizations and international bodies, Figure 2.6. These bodies have no executive role to play but are merely advisory. They have, however,

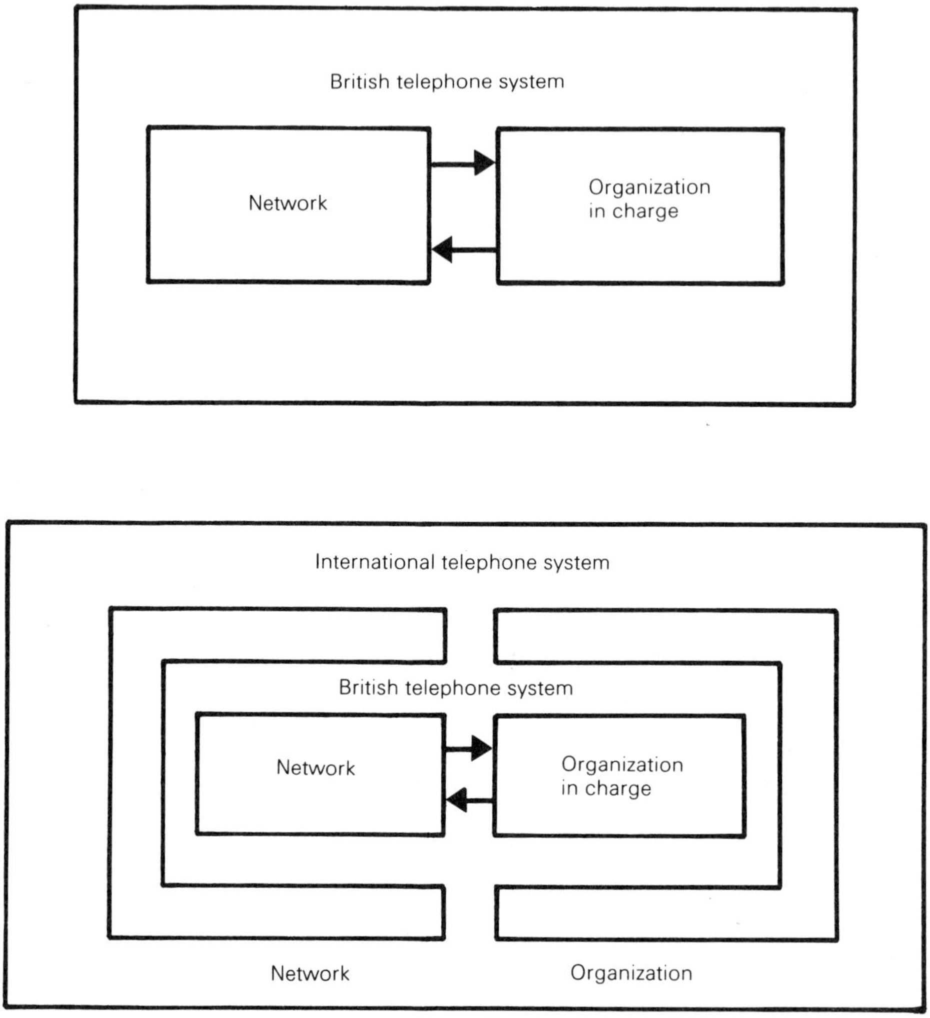

Figure 2.5

done many useful things of which the following are just two prime examples:

1. CCITT: this committee has established the protocol for interfaces on the switching operations for the network to take account of, among other things, the differences in the arrangement of numbers on the telephone dial, Figure 2.7. Clearly a Swede, when he dials a given

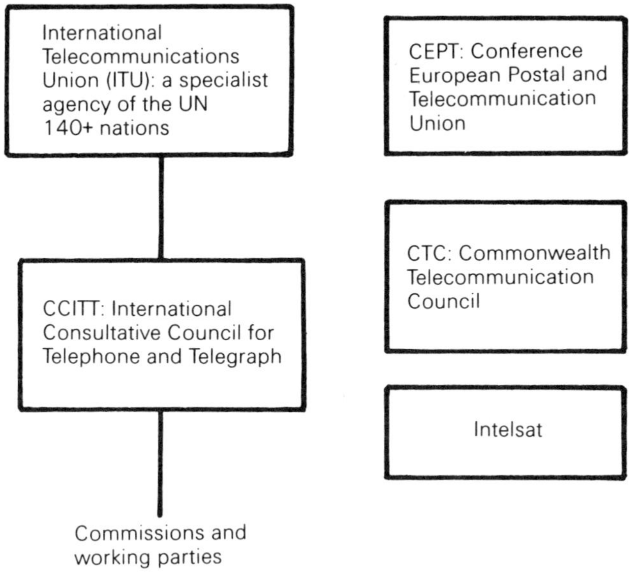

Figure 2.6

number, creates a different train of pulses than a subscriber in the United Kingdom or in New Zealand.

2. The world numbering plan: in anticipation of the time when all parts of the world would subscribe to the ITS, the appropriate international bodies devised a numbering plan whereby each country would be obtainable via a single digit being part of the complete number for a subscriber in that country. The following plan was approved:
   (a) 1 for North America;
   (b) 2 for Africa;
   (3) 3,4 for Europe (e.g. France – 33, West Germany – 49);
   (d) 5 for South America;
   (e) 6 for Australasia;
   (f) 7 for Russia;
   (g) 8,9 for Asia.

The major problem faced by the ITS is that of planning and in particular the provision of pathways. There are already a number of trans-Atlantic submarine cables, of which TAT5 is just one example. This cost £35 million to install and is a high-capacity, high-quality link, capable of supporting up to 825 simultaneous conversations. Its designed life is 20–25 years, although in practice it may last much longer. Following

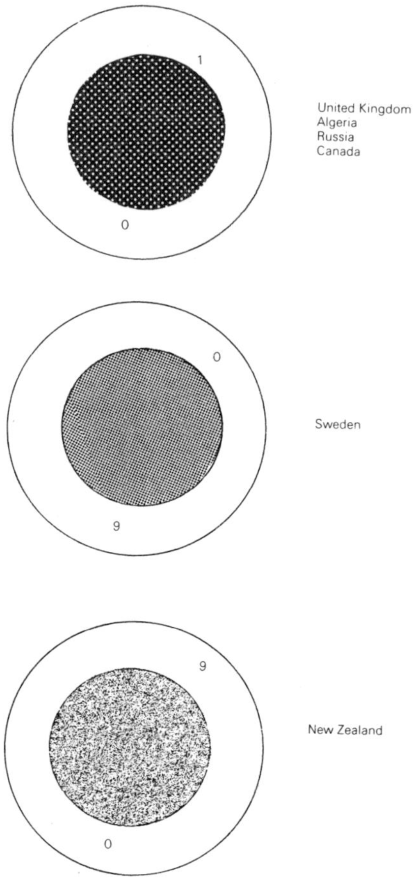

Figure 2.7

man's exploration into space, the communication system can now take advantage of satellite links and the INTELSAT system is a prime example here. These satellites rotate about the earth at a height of 22,000 miles above the equator and thus appear stationary over a given point on the earth's surface. They have a designed lifetime of 7 years, cost £5 million to build, £7 million to launch and their development costs exceed £30 million. There are at present eight satellites in all.

It is clear that the provision of these highly advanced technological communication links must be the responsibility of consortia, as appropriate, since the capital investment involved is beyond the means of any one nation. Communication is, in any event, a two-way process involving the origination and destination of calls, but in practice it is more multilateral than that. For example, the Commonwealth cable which links the United Kingdom, Canada, New Zealand, Australia, Hong Kong, Singapore and Malaysia was provided by these nations on a basis agreeable to each of them.

In spite of the large sums of money involved and the enormous complexity of a world-wide management system, the ITS continues to develop a more fully interconnected and cost-effective communications network.

The developing network will take advantage of the latest products of technology and gives improved quality and reliability of service. The Atlantic, typically London to New York, has five paths to offer so that if one is lost, 80 percent capacity is available for this particular demand. This is felt to be an adequate standard of reliability which is presently afforded to Europe and North America. The ITS's problem is to uphold this standard in satisfying the growing demand from the rest of the world for this vital large-scale system.

It is perhaps simpler to get an appreciation of the scale and complexity of this system than to perceive the system concepts which we must grasp. Having said that, however, a synopsis of statistics does little to convey a true sense of magnitude for this enormously complex and sophisticated system. It will have to suffice for the time being, however. As we go further into the book and examine the decision-making methods and design factors which enrich the interior of systems engineering we shall gain an underlying appreciation of the real scale and complexity of this type of system. We must turn, for the moment, to an enumeration of system concepts. It is clear that in one sense the ITS is an incorporation of sections of several other systems. It nevertheless has its own identity. In other words one can discuss this system in terms not only of its sections (or subsystems), but also of its own organization, resources and objectives. These are important system concepts, as too are the system's performance and its capacity to stimulate its own growth. In order to develop our understanding of these concepts further, we must now put a very important question, which is 'What do we really mean by the term "system"?' It is an unfortunate fact that the term has been much used and much abused to the point that it not only means very many different things to different people but it also is used to cover a multitude of sins. A formal description of the term 'system' is necessary in order to make sense of 'the systems approach' in general and of 'systems engineering' in particular.

2.4 SOME SYSTEM CONCEPTS

In order to get unanimous agreement on the definition of a term, one is obliged to keep it to the simplest of statements. In the case of the term 'system', we say that it is 'a collection of parts coordinated to accomplish a set of goals'. Though this appears not to say a great deal, it attracts little disagreement, and communicates the essence of what a system is supposed to be. We can now proceed to spell out in detail what the whole system is (the collection), what the environment in which it lives is, what its objective (the set of goals) is, and how this is supported by the activities of its parts.

There would seem to be five basic considerations to apply when thinking about the meaning of a system:

1. Total system objectives.
2. The system's environment.
3. The system's resources.
4. The parts of the system, their activities, goals and measures of performance.
5. The management of the system.

An examination of these will elucidate the basic system concepts we seek to establish. It should be stated at the outset that it is a non-trivial exercise to determine the precise nature of these considerations when applied to any given system. This is yet another reflection of the complexity of the systems engineer's task.

2.4.1 Objectives

There is a common difficulty in establishing objectives of any kind and that is to get at the heart of the matter. Once objectives become confused with viewpoints, we get different ideas of what the real objectives are, depending on whom we ask. It is like determining the real objectives of an individual person – in many, if not all cases, the truth remains hidden. In order to clarify the matter it is necessary to move from the vague statement of objectives to some precise and specific measures of performance of the overall system. In essence this is a 'score' that tells us how well the system is doing – better if higher, usually (unlike golf, of course). This simple idea has much appeal. If a single value can truly embody what the system is actually doing in terms of all of its activities, then the measure of performance says it all when it comes to judging the system through its objectives. Care needs to be exercised, however, in determining the correct measure of performance.

As an example, let us consider the losses of British merchant ships in the Mediterranean during World War II. To protect them from the attacks of German land-based dive-bombers, several hundreds of these ships were given anti-aircraft installations. These were expensive in terms of both guns and manpower. Therefore at the end of several months a re-evaluation of their effectiveness was made. It turned out that in only about 4 percent of the attacks were the enemy airplanes shot down. Since this was obviously a poor exchange ratio, it was tentatively decided to remove the anti-aircraft guns and crews from the merchant ships. However, the wrong measure of performance was being used. The one chosen (planes shot down) is obviously related to the question 'how can we destroy enemy aircraft most efficiently?' Clearly, there are better ways to do this than by putting anti-aircraft guns on merchant ships! The primary purpose of these guns was not to shoot down enemy aircraft but to protect friendly merchant vessels. If the guns succeeded in keeping the dive-bombers at a high altitude or forced them to maneuver evasively and if, as a result, the bombing aim was spoiled, the guns would have done their task. To determine the number of ships saved, it was necessary to compare the number of losses of ships protected by anti-aircraft guns with corresponding losses of unprotected ships. This is a measure of performance relative to the question 'how can the merchant ships be protected?' The figures indicated that only 10 percent of the protected ships had been sunk when attacked, whereas 25 percent of the unprotected ships had been sunk when attacked. When these facts were pointed out, it was realized that the guns offered a very fine return on the investment, and, correspondingly, a revised decision was made to continue this strategy.

Choosing the correct measure of performance in this case is equivalent to stating the problem correctly. For large-scale systems this, and not the solution of formulated problems, is almost without exception the more difficult exercise and it will not always be so straightforward a task to arrive at the correct measure of performance. Nevertheless this is an important concept in systems thinking. As many of the relevant consequences of the system activities as are possible to find are needed in the final determination of the system's correct measure of performance.

2.4.2 Environment

A second consideration is the environment of the system. This is what lies 'outside' the system which presupposes the existence of a boundary.

This in itself is a very important concept in systems thinking inasmuch as what is included as part of the system will clearly influence the thinking about the system. The idea of a boundary has been usefully described as 'a conceptual line drawn around the system such that it encloses all the important influences on the behavior of the system'. This is fine if, by implication, we now mean that the environment, that not enclosed by the boundary, makes up the things and people that do influence the behavior of the system, but which are 'fixed' or 'given' from the system's point of view. The definition of the boundary for any given system is quite crucial and its perception must be achieved by employing richer and more subtle means than a mere separation of self-evident system components from an 'obvious' external environment.

Often systems fail to perform properly simply because the managers have come to believe that some aspect of the world is outside the system and not subject to any control. In each case, therefore, of determining whether something belongs to the system's environment or not, we must ask 'can I do anything about it?' and 'does it matter relative to my objectives?' If the answer to the first question is 'no', but to the second is 'yes', then 'it' is in the environment. The concept of the system boundary underlines the fact that a systems approach is not an attempt to produce a model of something (some aspect of the real world). Rather it is a way of looking at and, for our purposes, deciding on a course of action for a particular something relative to answering a specific question. This concept is so crucial we shall return to it at the end of this chapter.

2.4.3 Resources

Our third consideration relates to the resources of the system. These are most definitely inside the system, i.e. enclosed by the system boundary. They are the means that the system uses to do its job and Jenkins (1969) has classified them, in alliterative style, as men, machines, money and materials. Resources, as opposed to the environment, are the things that the system can change and use to its own advantage. In looking and thinking about a system, we should pay attention not only to existing resources, but also to the manner in which resources can be increased. For example, by means of research and development in the case of hardware types of equipment, or by training and education of personnel or by various kinds of political activities which will increase the budget and investment potential. In fact, for many systems, a component that deals with the increase of resources may be the most important component of the system.

Resources are the general reservoir out of which the specific actions of the system can be shaped. The specific actions are taken by the parts, or subsystems. In order for our thinking about this important consideration not to become confused, we must distinguish between the traditional divisions employed, i.e. that of functional departments, and what we require to observe, namely the specific 'missions', 'jobs' or 'activities' of the system. Traditional divisions are not without vested interest from which one must expect to encounter resistance if thinking differently from this line. The questions that must be put are therefore 'is it worth it?' and 'why should it be done?' The real reason for the separation of the system is that it will yield the kind of information from which it can be inferred whether or not the system is operating properly, and what should be done next.

2.4.4 Subsystems

The ultimate aim of subsystem thinking is to discover those missions whose individual measures of performance are truly related to the measure of performance of the overall system. Ideally we should like the subsystem measures of performance to be in harmony with the overall system measure of performance. We shall find, however, that in the design of large-scale systems it can be very tricky and difficult to measure the performance of a subsystem. Moreover when designing systems from scratch, green-field design as it is sometimes called, it is important to realize that a system is greater than the sum of its parts. The system will exhibit features that are meaningful only at the systems level. They do not exist and are quite meaningless at the subsystem level. The system engineer ought to be capable of determining the emergent features of a system. This again is such a crucial issue we shall return to it in our summary at the close of this chapter.

2.4.5 Management

Finally we arrive at a consideration of the management of the system. This has to deal with the generation of the plans for the system, i.e. consideration of all of the things we have been discussing, the overall goals, the environment, the utilization of resources and the components. The management sets the subsystem goals, allocates the resources and controls the system performance.

Control, like measure of performance, system boundary, resource and subsystem, is another fundamental concept in systems thinking and it is necessary to investigate the modes of control and their effects on system management in order to understand it better. The simplest type of control is termed open-loop control. A predetermined plan is acted upon faithfully and no account is taken of the performance of the system during controlling action, it being tacitly assumed that all contingencies have been accounted for in the formulation of the plan. If the unforeseen occurs with open-loop control, there is no means of providing corrective or stabilizing action. A simple mechanical illustration of this type of control is the automatic washing machine. This has a simple programmed sequence which it obeys faithfully in open-loop fashion, proceeding, in most cases, through the three stages of soak, wash and rinse. A closer inspection of this piece of equipment will reveal, however, a different type of control, which is called closed-loop. For example, in the soak stage the tub is filled and once the desired water level is reached, the water inlet valve is shut off. If this were done in open-loop fashion, taking no account of the water level at any point in time, it would be necessary to plan for the water inlet to be open for a predetermined length of time. However, this assumes an availability of water, at a certain pressure, passing through a certain sized orifice giving, in total, a known flow rate. There may be conditions which cause a different flow rate to prevail, e.g. the unavailability of water, too high or too low a pressure, and so on.

These are the unforeseen circumstances against which open-loop control is powerless. By continually sensing the water-level, it is possible to feed this information back to the inlet waterline control valve and arrange for this to close at a desired water level. This is known as closed-loop control and requires a feedback loop to link the sensing of the output to the control of the input. In reality each stage of the automatic washing machine's cycle is governed by closed-loop control, the three stages being arranged as a whole in open-loop mode. Further to these two modes of control, there is another which is known as sampled-data or intermittent feedback control. This can be illustrated by the situation where the driver of a motor car aims to keep the vehicle at a constant speed. Clearly this cannot be achieved by a constant setting of the accelerator in the face of changing road conditions. Open-loop control is ruled out. The feedback loop consists of the speedometer, i.e. the sensor of the vehicle's speed, the driver and the accelerator. But the driver cannot be continuously watching the speedo since he must be watching the road. He can, in fact, only sample the speedo reading at intervals, hence the term intermittent feedback.

The exercise of the driver is not a difficult one, but one can imagine situations where the rate of sampling is important if effective control is to be maintained. This rate will vary depending on the conditions which are being encountered. During the sampling interval, events may happen about which nothing can be done until the subsequent sample is taken.

This is known as lag – formally expressed as the delay in getting information before being able to use it. When human beings are involved in systems, as they are in those systems in which we are chiefly interested, intermittent feedback is a common mode of control and lag is of great importance in the analysis of those systems.

2.4.6 Summary of concepts

We are now in a position to summarize the basic concepts that relate to our understanding of the term 'system' from the point of view of systems engineering. First, to function at all a system must have an objective. Our first concept is the system's measure of performance which tells us accurately and precisely the degree to which the system is fulfilling its objective. Secondly, we must know that which makes up the system and that which does not and yet which has an influence on the system's performance. This introduces the concept of a system boundary which delineates the environment of the system from the resources of the system. A system resource is our third basic concept which informs us to the wherewithal within the system for the attainment of the system's objectives. In essence a system resource is one of the following: men, money, machines or materials.

The concept of subsystems is fundamental in systems engineering. Systems may be broken down into subsystems, the subsystem detail depending on the problem being studied. This concept leads quite naturally to three further important properties of systems. First, a system is a complex grouping of human beings, hardware and organization (the last may include software). Secondly, the performance of each subsystem may interact with, and will probably be in mutual conflict with, those of the others. Finally, the system being studied will usually form part of a hierarchy of systems, i.e. the embedded property of a system. The final concept is that of control. A system should be designed in such a way that it is capable of achieving its overall objective in the best way possible. The management of the system must exercise control to ensure that the original plans are being carried out. This control also extends to the revision of those original plans since it can never be claimed that the system has been correctly defined in a once and for all manner. Control therefore further implies that management will receive information that tells it when its concept of the system is erroneous.

If these concepts have now been grasped, the reader might now care to answer the following question: 'what are the characteristics of the international telephone system and how can it be optimized?' It should be possible for the reader to make a sensible attempt at the first part of the question at least. The second part is much more open and really assumes a greater knowledge of systems engineering than that imparted by the book so far. The Greek origin of the word system is *systema*, which derives from *syn* meaning 'together' and *histemi* which means 'to set'. Typically, then, a system is a plan according to which things are connected into a whole. From the systems engineering point of view, a practical approach is called for to engineer a complex large-scale system from a collection of interacting subsystems so as to achieve an overall objective according to a plan that makes the whole system capable of achieving its objective in the best possible way. The question we now turn to is this: what ideas in systems engineering do we find that reveal the ways in which these systems concepts have been put to formative use?' Alternatively, what system methodologies do we find in current practice?

Classically the process of systems engineering is enacted in relation to each and every particular real-world problem as an individualized systems engineering study, and the study itself is accomplished traditionally by a systems engineering team. In the introduction of the subject of systems engineering to a beginner, therefore, these notions of 'study' and 'team' are quite fundamental. We shall now describe the constitution and employment of the subject we call systems engineering in terms of these notions and hence identify some prominent methodologies. Although the application of any systems approach to engineering is a relatively new development, no more than 50 years old, a body of people and of principles has become sufficiently well developed to be able to state that there are in essence at the present time two sharply distinct methodologies – though these are both closely connected to scientific method.

There is hard systems methodology, specific examples of which can be found in Goode and Machol (1957), and soft systems methodology, a relative newcomer to the scene; Checkland (1981). In both, the notions of 'study' and 'team' are perfectly valid. They address distinct problem types. In our discussion of these two methodologies we shall turn first to the traditional and reasonably well-established hard systems methodology, among whose contemporary proponents Finkelstein and McPherson are chief. To set the scene, however, we must define more precisely what we mean by a methodology.

2.5 METHODOLOGY

Methodology is not method. The latter derives from the Greek *methodos* which originally meant path (of one person pursuing another). It subsequently came to mean path (or way or route) in general and appears to have come to rest with the meaning 'mode of procedure, way or order of doing'; it may even mean system of classification though this merely serves to underline the poor esteem in which the word 'system' has come to be held. Nevertheless method connotes the meaning of exactness in technique or procedure that will take a defined start point and produce a desired end point. Methodology is a word that should convey a higher level of enterprise dealing more with a set of principles of method, established scientifically and/or systematically. A methodology will lack the precision of a technique but offer more guidance than a philosophy where the latter is ideological, broadly strategic and non-specific as to how the perceived 'what' is to be achieved.

To illustrate further: for the software expert, method may be an algorithm; methodology would then be structured programming; presumably philosophy then is the achievement of ends by means of symbolic processing and computational effort. Similarly to the hardware expert, method may be the electronic design of a counter circuit in an image processor; methodology would perhaps be the choice of a technology, e.g. transistor–transistor logic (TTL); the philosophy relates to means of hardware design in preference, say, to software programming a general-purpose digital computer.

Methodology in its simplest terms is the invention of methods. Though TTL was never intended to be a methodology originated specifically for the invention of methods, it nevertheless represents a well-defined framework in which the exploration of solutions to particular problems can conveniently proceed. Likewise, structured programming forces a discipline in software designers that facilitates the expeditious development of algorithms for well-defined problems. The point about levels of intellectual exercise matched to their respective theaters of operation is, however, well made.

The traditional sequence of events which orchestrates a methodology is as follows:

1. *Recognition* of problem.
2. *Definition* of problem.
3. *Action* to solve problem.
4. . . . Problem *solved*.

For the benefit of the pedant the final event signals the close of a methodology – the stopping criterion! Conversely what fires it into action, the first event, is a recognition of problem. What differentiates one methodology from another, are the intermediate events known as problem definition and problem solution. The broad class of problems for which problem definition is actually possible (and there are many which do not enter this class), are met at one by Ackoff's (1957) argument that 'all problems ultimately reduce to the evaluation of the efficiency of alternative means for a designated set of objectives.'

Leaving aside for the time being the objections that such evaluations, such definitions of efficiency and objectives, are realizable, we examine that body of knowledge which can be termed hard systems methodology, and in so doing look for evidence of systemicity, i.e. the amount of pure systems thinking that has gone into the development of that knowledge.

2.5.1 Hard systems methodology

An excellent account of the developments of these types of methodology is found in Checkland (1981) as he seeks to establish their essential goal-oriented nature in contrasting this with the softer systems methodology which is more concerned with the orchestration of debate centered on perceived problematical situations. The latter, Checkland argues, is more of the real world than the former at least so far as genuine problems are concerned as opposed to laboratory-sized experiments. Two prominent methodological types, systems engineering (Hall 1962, Machol 1965, Chestnut 1967, Chase 1974, Daenzer 1976 and Wymore 1976, and systems analysis (Hitch 1955, Quade 1963, Lee 1970, de Neufville and Stafford 1971 and Miles 1971 came into existence largely due to the incorporation of scientifically trained civilians in the planning of military operations during World War II. It is easy to see, for example, where operations research got its name. As it turned out, this became the precursor of systems analysis, thanks mainly to the sponsorship of the particular approach being taken in OR at the time, by politicians and well-endowed government agencies in United States.

Systems engineering and systems analysis can be differentiated only in terms of the relative emphasis on technology, and the processes which would use that technology, the former methodology clearly being preoccupied with the technology itself rather than its operational use. Hard systems thinking is fairly natural for design engineers whose role is to provide an efficient means of meeting a defined need. The design engineer's problem is a structured one, and the belief that real-world problems can be formulated in a structured way is the distinguished characteristic of all 'hard' systems thinking. Evidence of systemicity is confined largely to the use of systems concepts to achieve practical results – but this is to be expected. It is nonetheless clear that systems properties obtained gainful employment in the hands of systems engineers who were quick to identify

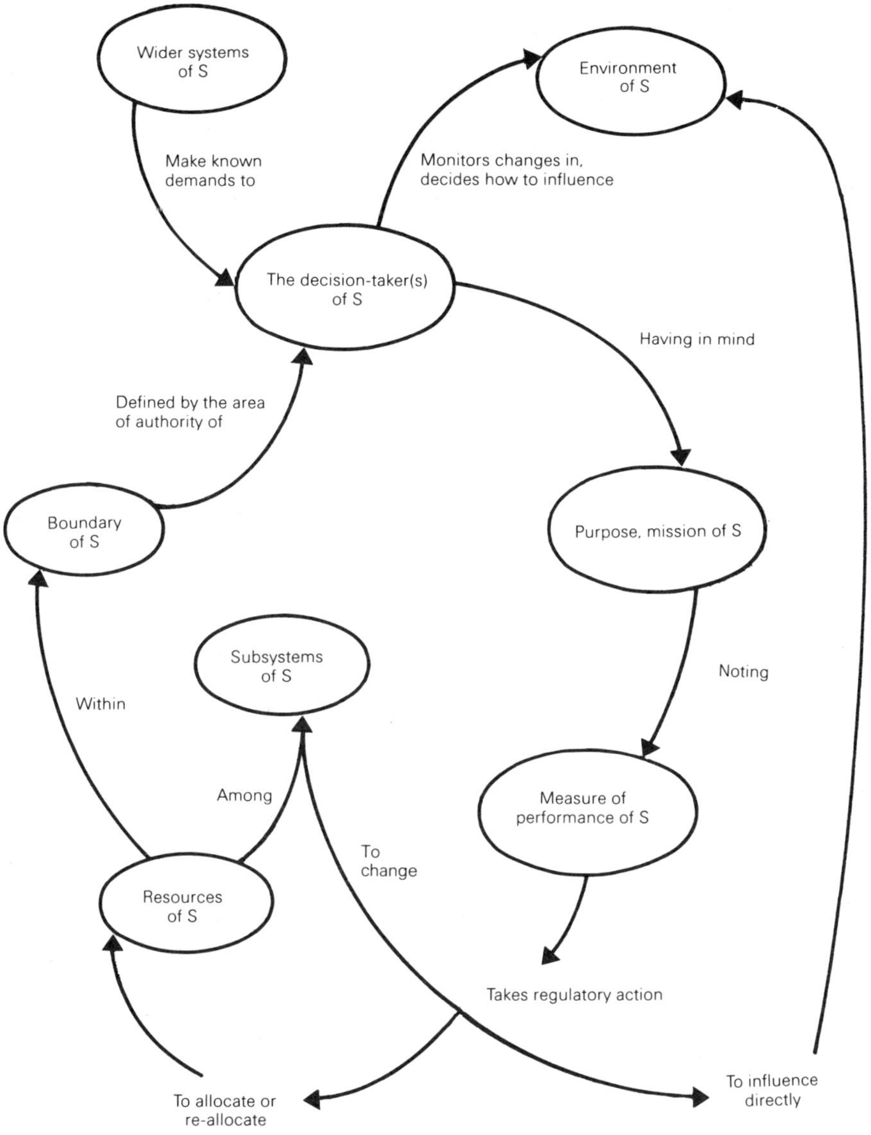

Figure 2.8

system boundaries, inputs and outputs (transfers across boundaries), components (subsystems), structure (interrelationship between subsystem), and the means of retaining the integrity of their defined systems (Figure 2.8). Sadly, however, attention to the coherency principle which makes it defensible to argue that the system is a system, must in retrospect have been somewhat sadly neglected as evidence of disagreeable system performance would tend to suggest.

Systemicity in the systems analysis methodologies tends to be illustrated by the manner in which these grew from initial tactical studies to encompass a broad class of less quantitative method, more oriented towards the analysis of broad strategic and policy questions.

In summary therefore the hard systems methodology proceeds along the following lines. It begins with a relatively well-defined problem whose owner is clear about the unsatisfactory state of affairs and the desired needs which are to be achieved. It proceeds to an analysis of objectives, measures of performance, system structure, the environment in which the system must operate and the requirements necessary to the accomplishment of the desired solution.

The next step is the design of the system itself which proceeds by quantitative model-building and simulation. Undoubtedly one very valuable credit to the systems movement in engineering is the identification for the benefit of all kinds of engineers in the various discipline of the system characteristics or phenomena which pervade throughout the profession. These are problems relating to traffic, reliability, competitive design and resource allocation and/or scheduling. This book, in its attempt to introduce a systems philosophy for engineering, has of necessity included a sample of the well-established methods and tools that have been developed for the treatment of just such phenomena. By this means the reader, having become adept in the use of these methods and tools, will feel a greater familiarity for the systems movement and may hopefully play a more active part in the gradual edification of a systems philosophy.

The next stage in the hard systems methodology is the optimization of the system design. Here again there is a well-established collection of methods, a sample of which is included in this book. The practical nature of the methodology obviously is recognizable from its final stage which is implementation. A description of this phase is beyond the scope of an introduction to the subject and the reader is advised to consult the excellent references cited at the end of this chapter.

2.5.2 Soft systems methodology

There has been for some 20 years or so now a school of thought at the University of Lancaster in the Department of Systems which believes that the real world or reality is more complex than hard systems methodology admits. Well-defined objectives, universally agreed problem definitions, recognizable alternatives, well-specified criteria for performance measurement – these are the parameters which are made of more elusive stuff in actual complex reality. Faults in hardware and software are, at least in principle, rigorously detectable. But who or what is to blame in the sinking of the *Titanic*, or the explosion of the shuttle Challenger? Designs in hardware and software are realizable by means of well-established tried and tested methods and techniques, but soft-systems thinkers will typically concentrate on questions such as 'how or by what means should British Coal be managed in order to achieve a maximum contribution to the UK economy?' – a very reasonable and technology-based policy problem. The characteristics of, say, a computer system can be reasonably well enumerated in terms of speed, reliability, power consumption, communications capacity and so on. Therefore enhancements to such a system may be unambiguously considered. But what are the characteristics of Gatwick airport and how can they be optimized? The question is far less straightforward, and essentially the reason why this is and why the previous two scenarios (dealing with technological failing and technological futures respectively) are similarly oblique is due to the presence of the human element within them.

Professor Peter Checkland, a leading exponent of soft systems methodology has developed schemes and principles which address problematical situations inherent in what he calls human activity systems. For the reader's benefit a summary of this methodology is given so that he may appreciate the broad spectrum of design which is becoming increasingly necessary to the professional engineer. If there is but one reason why a person with an interest in systems engineering should learn of the existence and nature of soft systems methodology, it is this: the rate of change of developments, which itself is increasing, in electronic system technology has forced complexity to become an issue in its own right. This is setting standards for society which becomes thereby even more demanding of the engineering profession as a whole. Any methodology which directly addresses complexity *per se* must be of immediate interest to those whose lives by dint of career choice will be concerned with nothing less than ever-increasing complexity.

Checkland began his researches with the intention of starting with the hard systems methodology and changing this appropriately to suit the new circumstances of the 'softer' problem. He subsequently found that major changes would occur as a result of the fact that problems could not be stated clearly and unambiguously. Therefore a methodology was required that provided a means of exploring that ambiguity and a means of orchestrating debate about change. Checkland found that change needed to be not only systemically desirable (i.e. appropriately consistent with the conclusions reached from applications of the methodology) but also culturally feasible (i.e. acceptance with the human activity system according to its peculiar norms, roles and values as meaningful and possible). An explanation of this methodology is facilitated by reference to Figure 2.9. From this it is clear that the entire process is mission-driven although precise definitions of the mission are indeed elusive and may never become precisely defined even after 'solutions' have been obtained, applied and the problematical situation removed!

The essence of the first two stages is to identify certain key roles such as: the client (one who commissions the study), the decision-taker (the system manager) and the problem-owner (the principal agent in providing the unstructured version in the problem situation). Following on from this, three essential features must be identified: structure, process and climate. These relate respectively, to the unvarying or slowly varying characteristics of the human activity system (e.g. power hierarchy), the dynamic characteristics (e.g. monitoring and reporting processes), and the interrelationship between structure and process. Checkland found that where structure and process are ill-matched (i.e. a bad climate), the real cause of the problem can be more readily identified and formulated. The overriding theme of these first two stages is to obtain the richest possible picture by eliciting as many relevant perspectives as may be found. The root definitions stage, stage 3 in Figure 2.9, seeks answers to the question 'what are the names of notional systems which from the analysis phase seem relevant to the problem?'

Given a problematical situation in a blood transfusion service a root definition might be: 'a transfer system concerned with locating a commodity (blood) in a particular place (the veins of potential donors), and removing it from there by some appropriate technology to a new location for storage.' A root definition should thus be a concise description of a human activity system which captures a particular view of it. In general, root definitions have the status of hypotheses concerning the eventual improvement of the problem situation by means of implemented changes which seem to both systems analyst and problem owners to be likely to be both feasible and desirable.

If the root definition is an account of what the system is, the conceptual model is an account of the activities which the system must do in order to be the system named in the definition, 3 of Figure 2.9. In no sense is it a description of any part of the real world: it is simply the structured set of activities which logic requires in a notional system which is to be that defined in the root definition. In other words Checkland is not modeling the real world – it is far too complex for that! He is drawing

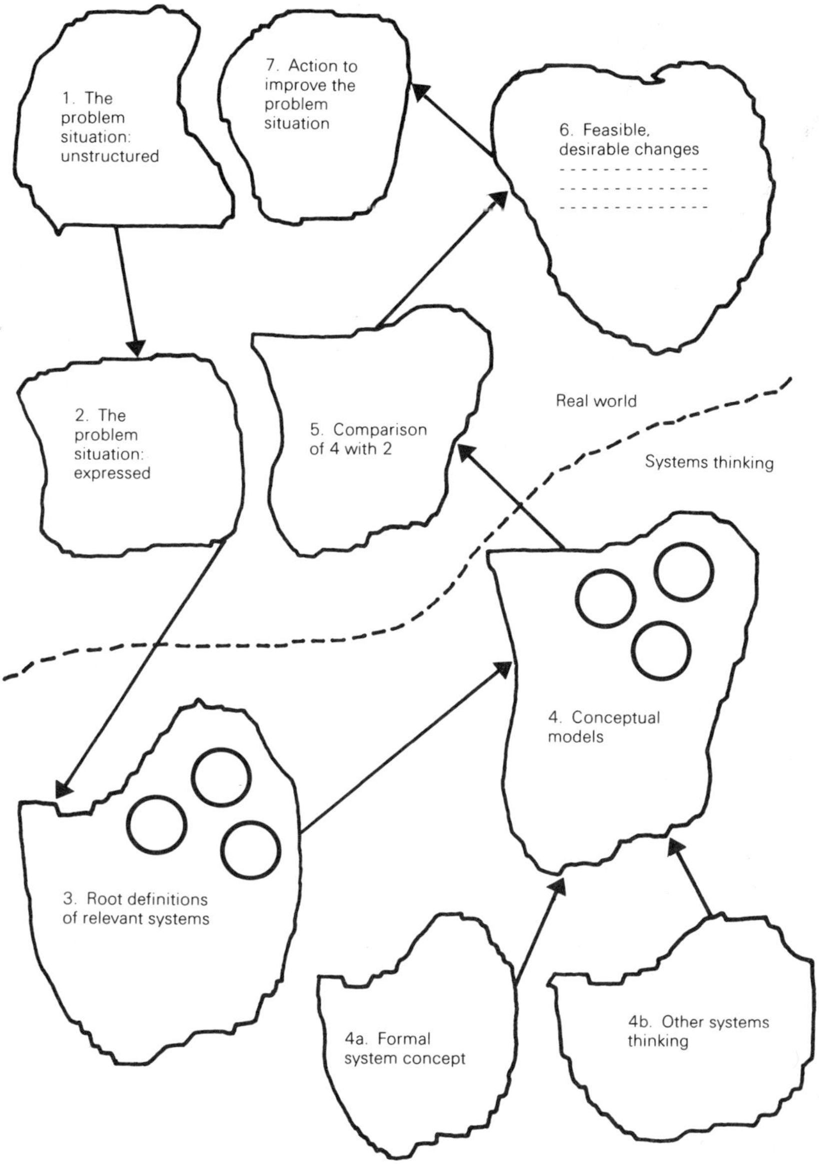
1. The problem situation: unstructured
7. Action to improve the problem situation
6. Feasible, desirable changes
2. The problem situation: expressed
5. Comparison of 4 with 2
Real world
Systems thinking
4. Conceptual models
3. Root definitions of relevant systems
4a. Formal system concept
4b. Other systems thinking

Figure 2.9

out the richest possible picture of a problematical situation and attempting to produce formal models relevant to that picture which can be compared with that particular aspect of the real world in order to generate feasible and desirable changes perceived to be improvements.

Because the conceptual model is a model of an activity system its elements are verbs: the modeling technique is to assemble the minimum list of verbs covering all necessary activities implied in the root definition and then to structure a sequence according to logic.

The resulting models are neither valid nor invalid – they are more or less defensible. The application of formal system concepts in the modeling technique makes it possible to develop models which are less deficient, less sloppy and more defensible. The comparison stage, 5 in Figure 2.9, embodies the basic systems hypothesis that systems concepts provide a means of teasing out the complexities of reality. One possibility is to use the conceptual models to open up debate about change, using them explicably as a means of putting questions to people involved in the problem situation. Another possibility is to compare what happened in the existing situation with what would have happened if the conceptual models had been implemented. But beware of making offensive recrimination of participants when considering their past performance. Here I slightly paraphrase a powerful statement made by Checkland in a form which I now commend to practicing engineers: 'comparison is the point at which intuitive perception of the problem is brought together with selected predications which assert an epistemologically deeper and more general account of the reality beneath surface appearances.'

In stages 6 and 7 in Figure 2.9 we note that changes brought about by hard systems methodology are represented nearly always by the creation and implementation of a system. This is less often the case for soft systems methodology where changes are likely to be of a much more model kind. They tend to be of three types: structural, procedural and attitude. The first two types of change obviously bear relation to the essential features identified in expressing the problem situation. The last is intended to include such things as changes in influence, and changes in the expectations which people have of the behavior appropriate to various roles as well as changes in the readiness to rate certain kinds of behavior as 'good' or 'bad' relative to others. In summary, to quote Checkland, this methodology is not 'a once-and-for-all approach to something sharply defined as "a problem" but as a general way of carrying out purposeful activity which gains from the power of some formal systems thinking but at the same time does not require individual human beings to behave as if they were rational automata.'

2.6 NEW DEFINITIONS

We are nearing the close of this chapter and the beginning of a section of the book which differs in tenor from that presented thus far. It is therefore prudent to summarize and establish quite clearly where we are in our introduction to systems engineering. Hopefully some key issues have become obvious by now and these can be succinctly stated as follows:

1. *Engineering* is about the real world, and the real world is about complexity! We must therefore always interpret the verb 'to engineer' as broadly as possible without ever losing sight of the abundant source of skills, techniques and knowledge relevant to technology itself. We seek as full a picture of engineering as we can, hoping to recognize the professionalism of the enterprise and make this known with due emphasis.
2. *Science* is essentially about differentiation (e.g. classification, discrimination, reductionism, analysis and labeling), whereas *systems* is essentially about integration (aggregation, coordination, cohesion, wholeness, synthesis and emergence). We would like to engineer systems that were defensible as such. We are still very much in a learning phase as to the meaning of these terms and how they can be applied, made operational, in a way that suits the action-oriented personality of the engineer. (A systems engineer will try to become systematically oriented but is likely to be no less of an engineer for it!)
3. *Design* is a fundamental component in the constitution of engineering and we hopefully have demonstrated its presence in both the hard and the soft systems methodologies. We would hope to maintain a broad spectrum of methodologies to which the systems engineer may refer and from which he may choose specific methodologies to deploy, for the benefit of supporting a systems approach to engineering. The design theme recurs through the book especially in the treatment of techniques (Chapters 4–7).
4. *Failures* are an inevitability, regardless of the richness of the design spectrum available to the engineer. These failures may range from the sudden and complete (i.e. catastrophic) to the gradual and partial (i.e. degeneracy). It behoves the engineer to take account of this as an inescapable consequence of the work that he does. Accordingly we, in this book, single out failure for peculiar treatment by identifying diagnostics as a key element in our problem-solving system, and by a full treatment of the subject of reliability in Chapter 4.

We recognize that the engineering profession as a whole has been lax at codifying the principles by which it has operated, with one result that it has an impoverished history, say in comparison with the law profession (significantly, law is based on precedence). The book seeks to make repairs to this by encouraging both the reading and writing of history and by making reference to the emergence of intelligent knowledge-based systems (IKBSs or expert systems) which require the elucidation of knowledge. The author is in no doubt that the engineering profession has much to offer and much to gain from this exciting development in computer applications, and these issues are discussed in Chapter 8.

Let us assume that a real-world problem can somehow or other be 'mapped' into a solution. Ackoff's maxim would be 'evaluate the alternatives', whereas Checkland would argue 'explore the ambiguities'. The philosophical framework proposed by this book places emphasis on a tension between differentiation (a scientific term) and definition (a systemic term). The definitions element takes responsibility for producing the systemic solution. Definition is a key therefore in generating solutions. Exploration and evaluation is of differences. For example: 'if it went wrong – what did we do differently from last time?' Or, 'if it is likely to change in the future, where will the differences lie?' Or, 'if I am to engage in design, how do my designs differ?' We now portray the essential defining terms with which a systems engineer will need to become skilled. They do not come with techniques. They exist to stimulate intellectual pursuit of systemic solutions to real-world problems. Only time will tell if they are desirable (superior to other concepts) and feasible (acceptable to problem solvers). There are three in all and each is introduced briefly in turn. A discussion is maintained of their application throughout the rest of the book as we turn first to techniques, and finally to a commentary on the efficacy of the entire systems engineering framework.

2.6.1 The first defining term: boundary constructor

If it were possible to construct system boundaries with unerring accuracy at all times we could define the perfect systems and solutions to everyone's pleasure (and amazement!). We cannot, but it helps to know so. But we nevertheless recognize the concept of the system boundary. After all is not problem definition really only the construction of a system boundary that embraces exactly the problem, the whole problem and nothing but the problem? When we know where boundaries are (around system, subsystem and all other things), cannot we then establish precisely the correct interfaces between them? We may get the boundaries wrong but the identification of which are to be the interfaces will have been correctly made. A boundary constructor is a most powerful device and worth its weight in gold. We have enumerated one boundary constructor, namely 'if it can be changed include it, if it cannot but it has an effect put it outside.' The question of where to put *it* inside (i.e. within or without a subsystem boundary is settled by the boundary constructor again, but this time the authority for change has moved to a different decision-taker.

This boundary constructor strongly relates to the design of a new system. It is deficient in relation to problems of technological failure (e.g. catastrophe investigation) or futures (e.g. determining the policies needed to integrate a nation's transportation infrastructures for the next 50 years). The search is on then for a boundary constructor. Attempts will be made to elucidate others throughout the book.

2.6.2 A second defining term: emergent feature

The more mature engineer will be one who has long since recognized that the system he designs not only does what he designed it for (novice engineers do not get this far!), but it also does other things that he never intended. Some of them are desirable. Some emergent features, for that is what they are, are less desirable. It could be argued that an emergent feature of a mass-produced television set system is to render sections of the viewing public intellectually bankrupt (sitting in front of the television starts early in life for many). Noise, air pollution and motorway madness are undesirable emergent features of the mass-producing automobile systems. Many years ago Professor Bill Gosling (1962) wrote 'The systems engineer must also be capable of predicting the emergent properties of the system, those properties that is, which are possessed by the system but not its parts.'

In the author's opinion we are no nearer today being able to achieve this than we were then. Even more tragically this laudable statement itself is no more impressed on engineers' minds and thinking than it was then. It is difficult to determine emergent features – but that should not be an excuse for not making the attempt. The point is made that there are some features which are meaningful only at the level of the system and perfectly meaningless below that level. A system, if it is to be defensible as such, must be greater than the sum of its parts. If we could be more confident in defining emergent properties then a system's greatness would owe less to notoriety and more to splendor. We initiate therefore a second search: for a definer of emergent features, with as much promise of success here as we might reasonably expect from our first search, but with no less hope.

2.6.3 The third and final defining term: hierarchical interdependence

In an earlier statement we said that systems are essentially about integration (aggregation, coordination, cohesion, wholeness, synthesis and emergence). If the boundary constructor defines the parts, and the emergent feature illustrates the greatness of the sum of the parts, then what do we have to say about how these parts are joined together? Questions arise on issues such as responsible agents, information flows, command sequences, reporting processes, interfaces, protocols and government. Structured programming is a technique or approach much in favor with computer experts at the present time. This is a means of achieving hierarchical organization of complex systems (a sophisticated divide-and-conquer technique). Similar approaches are rapidly having to be found in complex hardware design. From these methodological developments many lessons are being learned which point to ways of setting things together in a way that will achieve a system design that will convey the best and eliminate the worst. Of the three defining terms this last one looks to hold the greatest promise for the discovery of technique to govern hierarchical interdependence.

These three issues are to be kept in the forefront of the systems engineer's thinking if he is to develop systemic solutions as opposed merely to systematic solutions. It is the author's sincere hope that by giving due prominence to these defining terms within the framework set out to date, and by coupling this with the confident use of relevant design techniques, there will finally emerge in the reader's consciousness a keener sense of what constitutes complexity and how this is best confronted by a systems approach. The author recognizes that the reader is not being offered a technique to do engineering. No such thing exists nor could it ever. Hopefully the reader is being given more than just plain rhetoric. To some extent the engineering profession has lost sight of the glory that is engineering. (Why do so many of our most able pupils seek rewards in other careers?) The exciting developments in VLSI are enabling the true intellectual challenge that engineering offers to be drawn out. At the same time they shine a light on the deficiencies of pure scientific technique to cope with complexity.

The next section of this book describes a collection of techniques which are highly relevant to engineering. Their relevance is one reason for assembling them as such. But another important reason is the coherence they show collectively in reflecting these elements of design and diagnostics. Competence in the deployment of these techniques will give the reader a better understanding of the nature of engineering, and confidence to realize that more than technique is required.

2.7 EXERCISES

1. 'The history of man's survival is littered with changes: cultural, social and political. At every stage these changes have produced the legacies for successive generations. Man is no freer today of the consequences of his forebears' actions than they were of those of their own antecedents. In fact he is constrained to a greater degree because change and growth are considerably more powerful forces now.' Discuss.

2. Develop your own explanation for the rationale behind Figure 2.2, and, using a different frame of reference, produce another typology for large-scale systems.

3. Describe in your own terms, possibly using Section 2.3 as a model, *one* of the following complex systems:

(a) The Falklands 'war', 1982.
(b) The intercontinental ballistic missile (ICBM) system of the United States.
(c) The National Health Service in the United Kingdom.
(d) The Stock Exchange in London.
(e) Heathrow airport.

4. Choosing *one* of the systems given above, or some other suitable example, carry out *either* a 'hard' systems *or* 'soft' systems case study, making full use of the material in Sections 2.4 and 2.5. Examples offered are:

(a) 'Assuming that the air traffic through the London Terminal Management Area (LTMA) is to grow by 15 percent p.a., what steps must be taken in order to cope? For example, where should additional runway(s) be sited? What ground transportation infrastructure is needed? How can computers be of use?
(b) 'Assuming that motorway traffic in the United Kingdom is to increase by 15 percent p.a., what steps must be taken by the police in order to limit fatalities, e.g. how can the motoring public be alerted to dangers, educated on road safety, apprehended for law-breaking and penalized for the cause of disasters?'

5. Taking some situation in which you have some personal knowledge or a keen interest, develop your own ideas as to what constitutes: a boundary constructor; the means of defining emergent features in a system; and the policy for governing hierarchical interdependence. How generic are these ideas?

2.8 FURTHER READING

Ackoff, R.L. (1957) 'Towards a behavioural theory of communication', in W. Buckley (Ed.) *Modern Systems Research for the Behavioural Scientist*, Aldine (1968).

Blair, R.N. and Whitston, C.W. (1971) *Elements of Industrial Systems Engineering*, Prentice-Hall.
Bronowski, J. (1976) *The Ascent of Man*, BBC.
Buckley, W. (Ed.) (1968) *Modern Systems Research for the Behavioural Scientist*, Aldine.
Chase, W.P. (1974) *Management of Systems Engineering*, Wiley.
Checkland, P.B. (1981) *Systems Thinking, Systems Practice*, Wiley.
Chestnut, H. (1967) *Systems Engineering Methods*, Wiley.
Daenzer, W.F. (Ed.) (1976) *Systems Engineering*, Peter Hanstein.
de Neufville, R. and Stafford, J.H. (1971) *Systems Analysis for Engineers and Managers*, McGraw-Hill.
Goode, H.H. and Machol, R.E. (1957) *Systems Engineering*, McGraw-Hill.
Gosling, W. (1962) *The Design of Engineering Systems*, Heywood.
Hall, A.D. (1962) *A Methodology for Systems Engineering*, Van Nostrand.
Hitch, C.J. (1955) 'An appreciation of systems analysis' in S.L. Optner (Ed.) *Systems Analysis*, Penguin (1973).
Jenkins, G.M. (1969) 'The systems approach', *Journal of Systems Engineering*, **1** (1), pp. 1–21.
Lee, A.M. (1970) *Systems Analysis Frameworks*, Macmillan.
Machol, R.E. (1965) *Systems Engineering Handbook*, McGraw-Hill.
Miles, R.F. (Ed.) (1971) *Systems Concepts*, Wiley.
Optner, S.L. (Ed.) (1973) *Systems Analysis*, Penguin.
Quade, E. (1963) *Military Systems Analysis* in S.C. Optner (Ed.) *Systems Analysis*, Penguin (1973).
Sage, A.P. (1977) *Methodology for Large-scale Systems*, McGraw-Hill.
Shinners, S.M. (1976) *A Guide to Systems Engineering and Management*, Lexington Books.
Wymore, A.W. (1976) *Systems Engineering Methodology for Interdisciplinary Teams*, Wiley.

3 Basic probability theory

3.1 INTRODUCTION

Probability theory is a branch of mathematics, and as such it offers another means of developing mathematical models of physical situations. In the case of probability, the primary feature of the modeling technique is to deal with the uncertainty present in a physical situation. In reality, uncertainty is virtually always present in some form or other in every situation. For example, the definite arrival time of either people or transportation vehicles cannot be known for certain. One usually estimates what is likely to happen, i.e. the ETA is stated. Similarly, one can never be certain whether either a person or a system will fail, in any given period of time, to perform a stated operation. Though failure needs to be strictly defined, one is still faced with the uncertainty of its occurrence and again some form of estimate of its likelihood of occurrence must be made. Probability theory, and its adjunct mathematical statistics, combine to provide the modeler with a very powerful tool with which he can achieve this estimation. In this chapter we shall be concerned with the development of the ground rules for probability theory. We shall also be concerned with an examination of simple formulae which estimate the likelihood of events occurring within many familiar physical processes under some simplifying assumptions. In the two subsequent chapters we shall discuss the particular applications of probability theory to the very important analytical topics which recur continually in systems engineering studies, namely reliability and queues.

3.2 PROBABILITY NOTIONS

3.2.1 Sample space

A notion which is fundamental to the development of probability theory is that of sample space. This is the arrangement of all possible simple outcomes of a given physical process, where the occurrence of each of these is governed by chance. In probability terms the physical process is known as an experiment and the sample space is the set of experimental outcomes. If the experiment were that of tossing a coin, then we could write the sample space for this as S1, where

$$S1 = \{H, T\}$$

H denotes the simple outcome 'heads is the result of tossing the coin' and T denotes the simple outcome 'tails is the result of tossing the coin'.

Another experiment might be rolling a die which we can describe by the sample space S2, where

$$S2 = \{1, 2, 3, 4, 5, 6\}$$

where each number represents the mutually exclusive simple outcome denoting the appropriate number of spots on the upward face of the die following a single throw.

Notice how uncomplicated S2 is. The experiment could have been likened to S1 by having, say, S3, where

$$S3 = \{E, O\}$$

where E denotes an even number of spots on the upward face of the die following a single throw, and O denotes an odd number of spots being the result.

If we really wished to be obtuse, we could have had S4, where

$$S4 = \{L, G\}$$

where L denotes a number of spots less than four on the upward face of the die, and G denotes a number of spots greater than three.

The reason why we choose S2 in preference to S3 and S4 is because it alone details a mutually exclusive and collectively exhaustive set of irreducible simple outcomes which, in essence, defines the sample space for any experiment.

Notice, however, that E, O, L and G are all possible events in the experiment and they occur when one of the simple outcomes, which belongs to their particular event set, occurs. In other words, each event of an experiment can be described in terms of an event set of simple outcomes which is necessarily a subset of the sample space. The event set relates to the event in exactly the same manner as the sample space, itself a set, relates to the experiment, as a whole. We could write, therefore:

$$E = \{2, 4, 6\}; \qquad O = \{1, 3, 5\};$$
$$G = \{4, 5, 6\}; \qquad L = \{1, 2, 3\};$$

and in set theory notation $S2 = E \cup O = G \cup L$. Notice $E \cap O$ and $G \cap L = \Phi$, the empty set.

3.2.2 Axioms

The second important notion in the development of probability theory is the probability axiom, and, generally speaking, three are surmised. These are as follows:

1. $p(A)$ is defined as the probability of the event (or event set) A and is non-negative, i.e. $p(A) \geqslant 0$.
2. $p(A \cup B) = p(A) + p(B)$ when A and B are mutually exclusive events and $A \cup B$ denotes the union of sets A and B. Note $A \cap B = \Phi$.
3. $p(S)$ is defined as the probability of the experiment itself and is always equal to 1. This is the certainty probability value. Note that it follows that $p(\Phi) = 0$.

The reader should now deduce for himself that for a fair coin, i.e. equal likelihood (or equal probability) of the events H and T, then $p(H) = p(T) = 1/2$, the intuitive result. It should also be possible to deduce that from the second experiment, if the die is fair, $p(E) = p(O) = p(G) = p(L) = 1/2$, and $p(1) = p(2) = p(3) = p(4) = p(5) = p(6) = 1/6$.

3.2.3 Probability laws

The third important notion is that of the probability law and there are two such laws of great significance which we shall now derive. The first is the probability law of addition for non-mutually exclusive events. Referring to Figure 3.1, we note that:

$$A \cap B \neq \Phi$$

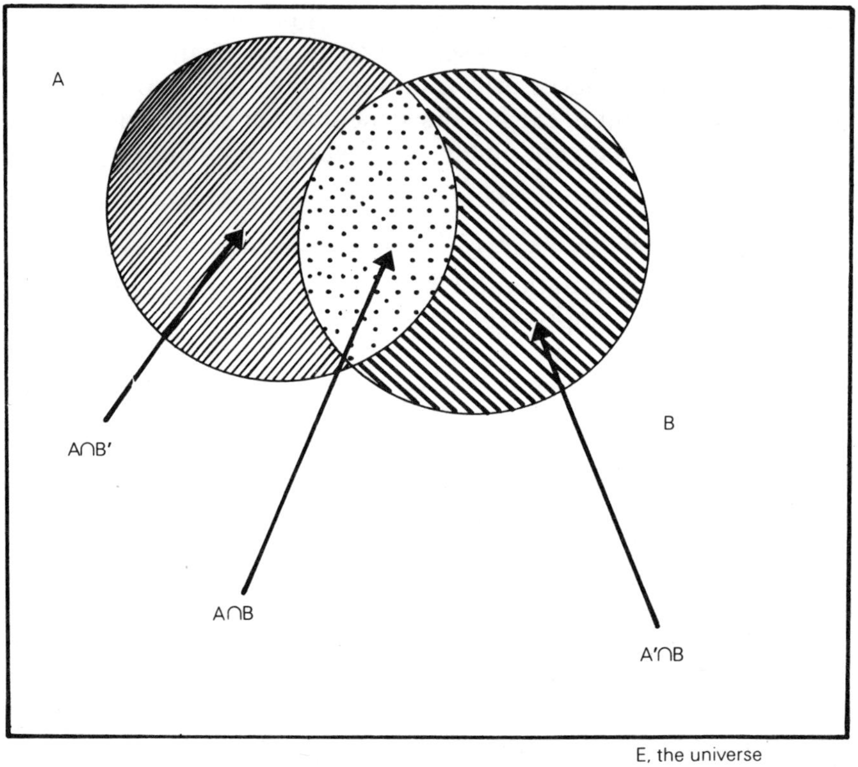

Figure 3.1

We adopt the usual notation of set theory and define the complement of the set A to be A′ (not A) where it is understood that

$$A \cup A' = E \text{ and } A \cap A' = \Phi$$

Then

$$A \cup B = (A \cap B') \cup B$$

but

$$(A \cap B') \cap B = \Phi$$

so applying axiom (2), we get

$$p(\mathrm{A} \cup \mathrm{B}) = p(\mathrm{A} \cap \mathrm{B}') + p(\mathrm{B}) \tag{3.1}$$

Now

$$\mathrm{A} = (\mathrm{A} \cap \mathrm{B}') \cup (\mathrm{A} \cap \mathrm{B})$$

but

$$(\mathrm{A} \cap \mathrm{B}') \cap (\mathrm{A} \cap \mathrm{B}) = \Phi$$

so applying axiom (2) again, we get

$$p(\mathrm{A}) = p(\mathrm{A} \cap \mathrm{B}') + p(\mathrm{A} \cap \mathrm{B}) \tag{3.2}$$

Applying equations (3.1) and (3.2) simultaneously gives

$$p(\mathrm{A} \cup \mathrm{B}) = p(\mathrm{A}) + p(\mathrm{B}) - p(\mathrm{A} \cap \mathrm{B}) \tag{3.3}$$

which is the desired result. Notice that this reduces to the second axiom when A and B are mutually exclusive.

The reader should now calculate the results $p(\mathrm{E} \cup \mathrm{G})$ and $p(\mathrm{O} \cup \mathrm{L})$ for S2 above.

Our second probability law is the product law for independent events and leads us to a clearer understanding of the term $p(\mathrm{A} \cap \mathrm{B})$ in equation (3.3).

With reference to our second experiment, S2, it is possible to ask the question 'given that we obtain an even number, what is the likelihood of obtaining a two?' Equally we could ask 'given that we obtain a number less than four, what is the likelihood of obtaining an even number?' This type of question is expressed in its general form as follows. On the basis that some event is known to have happened, what is the dependent likelihood that another event has happened as a result? This dependent likelihood or dependent probability is of key interest in the general development of probability theory. It is written as $p(\mathrm{B}/\mathrm{A})$ meaning the (dependent) probability of B occurring given that A has occurred. Conversely $p(\mathrm{A}/\mathrm{B})$ means the (dependent) probability of A occurring given that B has occurred. Clearly both of these terms are related to the term that expresses the probability of the simultaneous occurrence of both A and B which is given by $p(\mathrm{A} \cap \mathrm{B})$. What, expressly, then is the relationship between these three terms? To answer this question we really need to refer to the frequency model of probability which was the original means in the development of the theory. The notion here is that in the repeated application of an experiment, the likelihood of a given outcome is measured by the ratio of the number of times which this outcome occurs to the number of repeated trials of the experiment, i.e.

$$p(\mathrm{A}) \approx N_{\mathrm{A}}/N$$

where N is the number of trials and N_{A} the number of occurrences of the event A in that number of trials. It should be obvious then that $p(\mathrm{A} \cap \mathrm{B}) \approx N_{\mathrm{AB}}/N$ where N_{AB} is the number of simultaneous occurrences of both A and B. A simple manipulation of this gives:

$$p(\mathrm{A} \cap \mathrm{B}) \approx N_{\mathrm{AB}}/N = N_{\mathrm{AB}}/N_{\mathrm{A}} \cdot N_{\mathrm{A}}/N = p(\mathrm{A}) \cdot N_{\mathrm{AB}}/N_{\mathrm{A}}$$

What do we infer as the ratio $N_{\mathrm{AB}}/N_{\mathrm{A}}$ however? If we treated A as the sample space in which it was possible but not certain for B to occur, then this ratio expresses the likelihood of B occurring on the clear basis that A has already occurred, i.e. $p(\mathrm{B}/\mathrm{A})$. Therefore:

$$p(\mathrm{A} \cap \mathrm{B}) = p(\mathrm{B}/\mathrm{A}) \cdot p(\mathrm{A}) \tag{3.4}$$

We find, similarly, that:

$$p(\mathrm{A} \cap \mathrm{B}) = p(\mathrm{A}/\mathrm{B}) \cdot p(\mathrm{B}) \tag{3.5}$$

We can now answer the two questions posed above in relation to S2, i.e. to find $p(2/\mathrm{E})$ and $p(\mathrm{E}/\mathrm{L})$. Notice that $p(\mathrm{E}/2) = 1$, and we know $p(2) = 1/6$, therefore:

$$p(2 \cap \mathrm{E}) = p(2)\, p(\mathrm{E}/2) = 1/6$$

Therefore

$$p(2 \cap \mathrm{E}) = 1/6 = p(\mathrm{E})\, p(2/\mathrm{E})$$

But

$$p(\mathrm{E}) = 1/2$$

Therefore

$$p(2/\mathrm{E}) = (1/6)/(1/2) = 1/3$$

the obvious result.

Also since $\mathrm{E} = \{2, 4, 6\}$ and $\mathrm{L} = \{1, 2, 3\}$, then

$$\mathrm{E} \cap \mathrm{L} = \{2\}$$

Therefore

$$p(\mathrm{E} \cap \mathrm{L}) = 1/6$$

Whence

$$p(\mathrm{E}/\mathrm{L}) = p(\mathrm{E} \cap \mathrm{L})/p(\mathrm{L}) = (1/6)/(1/2) = 1/3$$

again the obvious result.

An important law in probability theory is the law of independent events, when the probability that an event occurs is in no way dependent on the occurrence of another event, i.e. independence in the occurrence of events. This law states that the probability of the simultaneous occurrence of independent events is the product of the event probabilities. If, therefore, we rolled two dice, the probability of getting two even numbers is the product of the two respective probabilities of getting an even number for each die, i.e.

$$p(E1 \cap E2) = p(E1) \cdot p(E2) \tag{3.6}$$

where E1 denotes the event an even number obtained from the single throw of die 1 and E2 refers to the same event for die 2. The reader should now show that the probability of throwing a total of seven on the single roll of a pair of dice is 1/6 using the probability law of independent events and axiom 2. What we have done so far is summarized in Figure 3.2.

An important third law in probability theory which follows quite naturally from what we have seen so far is the conditional law of probability. If an event A may occur dependent on the outcome of any of several mutually exclusive events, say Bi, then the probability of the event occurring p(A) is given by equation (3.7):

$$p(A) = \Sigma_{i=1,n}\, p(Bi) \cdot p(A/Bi) \tag{3.7}$$

In the simple case of $n = 2$ this resolves as:

$$p(A) = p(B1)\ p(A/B1) + p(B2)\ p(A/B2) \tag{3.8}$$

If derived, the Bi can be likened to certain causes, which may or may not occur, and A an effect which may occur as a consequence of the causes formerly occurring. This leads us to ask the question: 'given that effect A has been monitored, what is the likelihood that one cause in particular gave rise to this effect?' This probability, in the case of cause Bj, is expressed as $p(Bj/A)$. This question is answered by referring once again to the simultaneous occurrence of dependent events, i.e. to the event $A \cap Bj$.

We know that

$$p(A \cap Bj) = p(A)\ p(Bj/A) = p(Bj)\ p(A/Bj)$$

Whence

$$p(Bj/A) = p(Bj)\ p(A/Bj)/p(A) \tag{3.9}$$

This is an important result and is a simple yet effective illustration of

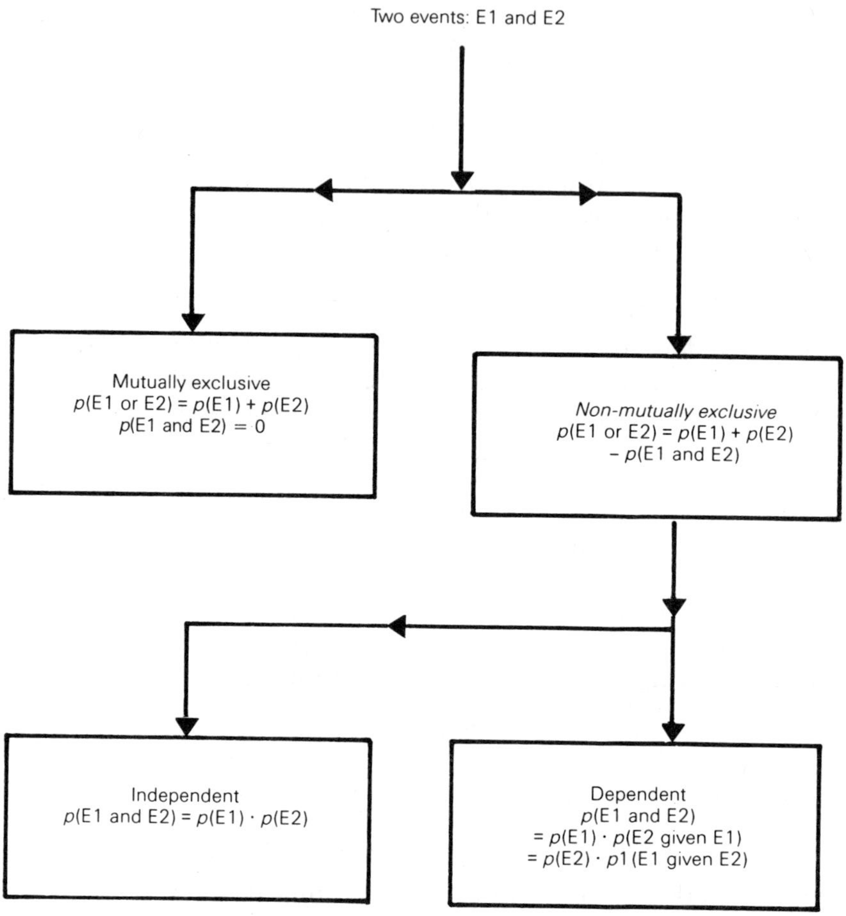

Figure 3.2

Baye's theorem, which defines the a posteriori (or reflexive) probabilities that certain events (causes) gave rise to particular dependent events (effects) that are known to have happened. If in general there are n causes and m effects, i.e. Ak, $k = 1, 2, \ldots, N$, then the probability that Bj caused the given effect Ai to occur is:

$$p(\mathrm{B}j/\mathrm{A}i) = (p(\mathrm{B}j)p(\mathrm{A}i/\mathrm{B}j))/\Sigma_{k=1,N}\, p(\mathrm{B}k)\ p(\mathrm{A}i/\mathrm{B}k) \qquad (3.10)$$

To summarize nomenclature: the $p(\mathrm{A}i)$ and $p(\mathrm{B}j)$ are known as a priori probabilities; the $p(\mathrm{A}i/\mathrm{B}j)$ are known as dependent, conditional or transitional probabilities; the $p(\mathrm{B}j/\mathrm{A}i)$ are known as reflexive or a posteriori probabilities, it being assumed that the Bj are *causes* and the Ai are *effects*. Let us now consider two worked examples to illustrate the application of these probability laws to some suitably imaginary situations.

Example 3.1

A manned rocket vehicle is estimated to have a reliability (probability of mission success) of 0.81 on the first launch. If the vehicle fails there is a probability of 0.05 of a catastrophic explosion, in which case the abort system cannot be used. If an abort can be attempted, the abort system reliability is 0.9. Calculate the probability that the crew survives the first launch.

Let: E1 be mission success event
E2 be non-catastrophic failure event
E3 be successful abort event
E4 be crew survives event

Then:

$$p(\mathrm{E1}) = 0.81$$

and

$$p(\mathrm{E1'}) = 1 - p(\mathrm{E1}) = 0.19$$
$$p(\mathrm{E2'}) = p(\mathrm{E2'/E1'}) \cdot p(\mathrm{E1'}) = 0.05 \times 0.19 = 0.0095$$

Similarly:

$$p(\mathrm{E2}) = p(\mathrm{E2/E1'}) \cdot p(\mathrm{E1'}) = 0.1805$$

Also

$$p(\mathrm{E3}) = p(\mathrm{E3/E2}) \cdot p(\mathrm{E2}) = 0.9 \times 0.1805 = 0.16245$$

and

$$p(\mathrm{E4}) = p(\mathrm{E1}) + p(\mathrm{E3}) = 0.81 + 0.16245 = 0.97245$$

From this, notice the following:

1. The crew survives either if the mission is a success, or, if not a success, there is not a catastrophe and the abort system works.
2. A non-catastrophe is dependent first on a non-successful mission.
3. A successful abort is dependent first on a non-catastrophe.

Note: $p(\mathrm{E4}) + p(\mathrm{E2}) \neq 1$. Why?

Example 3.2

An inventory of transistors consists of 60 percent from manufacturer X and 40 percent from manufacturer Y. The proportion defective for X is 0.01 (i.e. probability that a transistor made by X is defective is 0.01), and for Y is 0.02. A transistor is drawn at random and found to be OK. What are the respective probabilities that first, it came from X and secondly, that it came from Y?

Let X denote the event – transistor made by X
Y denote the event – transistor made by Y
D denote the event – transistor defective
G denote the event – transistor good

Now for the data: $p(X) = 0.6$, $p(Y) = 0.4$, on the basis of a random sample. The conditional probabilities which are given are

$$p(D/X) = 0.01 \text{ and } p(D/Y) = 0.02$$

From this we may conclude that $p(G/X) = 0.99$ and $p(G/Y) = 0.98$. Using the conditional probability law gives:

$$\begin{aligned} p(D) &= p(X)\,p(D/X) + p(Y)p(D/Y) \\ &= 0.6 \times 0.01 + 0.4 \times 0.02 \\ &= 0.006 + 0.008 = 0.014 \end{aligned}$$

Similarly $p(G) = 0.986$. Now we are given the information that the transistor is good and asked first the probability it came from X and, secondly, the probability it came from Y, i.e. $p(X/G)$ and $p(Y/G)$.

Using Baye's theorem:

$$\begin{aligned} p(X/G) &= p(G/X)\,p(X)/p(G) \\ &= (0.99 \times 0.6)/0.986 = 0.6024 \end{aligned}$$

and

$$\begin{aligned} p(Y/G) &= p(G/Y)\,p(Y)/p(G) \\ &= (0.98 \times 0.4)/0.986 = 0.3976 \end{aligned}$$

Notice these results sum to unity, i.e. the transistor must have come from either X or Y, but notice also that the information 'transistor is good' favors X because X makes slightly better transistors. The reader should now calculate the a posteriori probabilities that a transistor found to be defective may have come either from X or from Y.

3.2.4 Random variables

The fourth important notion in probability theory is that of the random variable. This is a rule for expressing the result of an experiment as a number, e.g. in the experiment, tossing a fair coin, the random variable can be the rule that assigns a value 0 to the result if heads is the outcome, and 1 to the result if tails is the outcome. This coding operation is often quite useful, e.g. where it is necessary to say the experiment was successful or unsuccessful. Other simple random variables are the temperature of a chemical process, the resistance of a thermocouple, the height of males in Europe over 21 years of age, number of actual aircraft arrivals at Heathrow airport in an hour, etc.

It should be remembered that the random variable is a rule and therefore a function for determining a numerical value which expresses the result of an experiment whose outcomes are governed by chance.

We shall denote the random variable by an upper case letter, e.g. X, and the values which it may assume (called variates) by lower case letters, e.g.

X is the height of males over 21 years in Europe
$x1 = 1.78\,\text{m}$ $\quad$ xi are variates of X
$x2 = 1.88\,\text{m}$ etc.

Random variables, real valued functions defined on the sample space, may be either discrete or continuous. Clearly if the random variable is continuous there are an infinite number of variates for it, but this is possible even if the random variable is discrete. We shall speak of a probability rule for the random variable that expresses the probabilities that the random variable will assume certain values. The distinction between discrete and continuous is made here in that for a discrete random variable we can speak of $p(X = x)$ whereas for a continuous random variable we cannot speak of $p(X = x)$ but only $p(x_a < x < x_b)$.

The rules which assign probabilities to the variates for a random variable will be called probability distributions, to which we will turn in a moment, but we now briefly mention the expected value function defined for random variables.

Discrete: $E(X)$: $\Sigma_{i=1}\, xi\; p(xi)$
Continuous: $E(X) = \int xp(x)\,dx$

The expected value function is not just a mathematical nicety invented for its own sake at the fancy of mathematicians. It is the means of defining important statistics, two chief examples of which are mean and variance. By definition, the mean of a random variable is simply its expected value. A more complicated statistic is the variance, which is the difference of two terms. The first is the expected value of the square of the random variable and the second is the square of the mean of that random variable, i.e.

$$\text{variance}\ (X) = \mathrm{E}(X^2) - [\mathrm{E}(X)]^2$$

The subject of mathematical statistics makes great use of these and other terms. In the later chapters discussing the application of probability theory to the analysis of the problematical aspects of systems engineering studies, we shall see how in particular the mean and variance of random variables relate to such terms as failure rate characteristics, and arrival and service patterns.

3.3 PROBABILITY DISTRIBUTION LAWS

3.3.1 Binomial probability distribution

Let us now, however, consider a simple problem which elaborates the concept of the random variable. A certain manufacturing process is found to produce some items which are defective in terms of the adopted standards of quality. These defective items appear randomly but, on average, one manufactured item in a hundred is defective. The manufacturer wishes to look on occasions at a sample, selected at random, of, say, five items. We want to know what the likelihood is of finding fewer than two defective items in a given sample. On examination, there are five ways in which one defective item may appear in a sample of size five and only one way in which no defective items may appear. They are as follows:

BGGGG	GGGBG
GBGGG	GGGGB
GGBGG	GGGGG

where B denotes a defective item and G a non-defective item.

Each of these ways is mutually exclusive to the others. Collectively they represent the event of interest, i.e. fewer than two defective items in a sample of size five. If we denote this event as A, then $p(\mathrm{A})$ is given by the sum of the probabilities of these six ways. Let us look at the last of these ways in which there are no defective items. This represents the simultaneous occurrence of five independent events in which the event probability $p(\mathrm{G})$ is constant and known to be 0.99 ($= 1 - p(\mathrm{B})$). The probability of this 'sixth' way is therefore $(0.99)^5$ (= 0.951). By similar reasoning the probability for each of the first five ways has the same value and this is given by the product of the event probabilities, i.e. $(0.99)^4 \cdot (0.01)$ (= 0.00961). The desired answer therefore is approximately 0.99902.

We can interpret this result as follows. Knowing that on average one in a hundred items are defective, then, taking samples of size five, on average, slightly less than one in a thousand samples will have more than a single defective item. If the manufacturer found, by experience, a contrary result he might suspect that the average of one defective item in a hundred was no longer correct. This in itself appears to provide useful information, but the situation is imaginary and we are for the moment more concerned with what knowledge can be made from this kind of probability analysis. Let us try to formalize this. First for a sample size n we can define the random variable X which is a rule for determining the number of defective items in that sample.

The values of X may be anything between 0 and n inclusive, and X is defined as a real-valued discrete function for the experiment 'draw a sample of items of size n at random from the said manufacturing process'. We now need to determine the probabilities which govern this random variable assuming its values (variates). Such an arrangement of probabilities for a random variable is known as a probability distribution. For this particular random variable the distribution is easily found. Let us assume a particular variate k, i.e. that in a sample of n items, k are found to be defective. The number of ways in which this can happen totals

$$nCk = n!/(k!(n-k)!)$$

where nCk means choose k from n.

The probability of each of these ways has the same value and it is the product of the simultaneous occurrence of k defectives and of $(n-k)$ non-defectives, i.e. $(p)^k(1-p)^{n-k}$, where p is the constant and independent probability that any item selected at random from this process is defective. This is sometimes known as the proportion defective. The probability therefore that X assumes the value of k is given by:

$$p(X = k) = n!/(k!(n-k)!)(p)^k(1-p)^{n-k} \quad (3.11)$$

This is known as the *binomial probability distribution law* and applies to discrete random variables under the following conditions:

1. The experiment must consist of a finite number of discrete outcomes.

2. Each outcome must be defined in terms of one of only two possible mutually exclusive states.
3. The probability that an outcome assumes a given state is constant and independent from outcome to outcome.

The reader should now attempt the following problem. Assume in the manufacture of widgets, a proportion defective of 1 percent. What is the probability that in a random sample of a hundred, fewer than two will be defective? (Answer: 0.73576 approximately.) In some cases a certain amount of care is required before seeing if the binomial distribution actually applies to a situation which at first glance calls for its use. The following example will illustrate.

Example 3.3

Suppose that the probability of a person being killed on any given airplane flight is 10^{-6}. What is the probability that a person, the same one, will be killed in attempting 10^6 such flights? We must immediately dismiss one as an answer. It is by no means certain that the person would not survive one million such flights. We should, however, expect that on average, there would be one airplane disaster per million flights. Can we use the binomial here? For example, can we make the assumptions which are necessary for the application of the binomial, in which case we would make $p = 10^{-6}$ and then seek $p(X = 1)$, where X is the random variable determining the number of deaths of the person in a sample of one million flights? The answer to this is 'no'. It is meaningless to talk of $p(X = k)$ where $k > 1$. In other words, if a death occurs before the million flights are up, the remainder will not be undertaken and so the application of the binomial is not appropriate. We can, however, use it rather neatly, if we talk in terms of survival, i.e. zero deaths. This probability is predicted by the binomial, when $p(X = 0)$ gives the answer. The converse, not surviving, must therefore be given by $1 - p(X = 0)$. The probability value we require therefore is

$$1 - (1 - 10^{-6})\hat{}10^6 \approx \exp^{-1} \approx 0.36$$

i.e. 64 percent chance of a fatality in 10^6 flights.

3.3.2 Poisson probability distributions

We consider now a second probability distribution law, called Poisson, which may be derived from the binomial but for which different assumptions apply. It is evident in the case of the binomial that not only can outcomes of one type be counted but so also can the outcomes of the other type, e.g. defectives and non-defectives.

There are many examples of situations where outcomes can appear and be counted, and where the non-appearance of the outcome is not countable but is, theoretically at least, an infinite number. In any given period of time, the number of failures of a system, the number of arrival of people or transportation means, e.g. cars, trains, airplanes, or the number of assaults by an enemy can all be counted. What cannot be counted is the number of non-events relating to such, e.g. a non-arrival. Similarly in a piece of wire or glass the number of defects can be counted but the non-defects cannot. Theoretically an unlimited number of opportunities exist in a space or time continuum for such events as defects, failures, arrivals and assaults. The character of dichotomy which is fundamental to the binomial is once again exhibited in this situation but the finiteness of the random variable is not.

Let us assume that for a continuum of given length there may appear at random a certain event and that on average, for this given length, there are λ such events, which is approximately constant. Assume now this continuum to be divided into n equal subintervals and let n be so large that it is impossible for more than one event to appear in any one subinterval. Let the probability that an event will occur in a subinterval be p. Then the probability for its non-appearance is $1 - p$. By definition of the average we establish the relation $pn = \lambda$ or

$$p = \lambda/n$$

If n is sufficiently large so as to prohibit the appearance of more than one event, then the probability of the appearance of an event in any subinterval of corresponding size is λ/n. Since conceptually it takes only an instant in time for an event to occur, or an infinitessimal amount of space for a defect to appear, then the value of n must tend to infinity. Under these circumstances we have a model of the physical process to which the binomial probability distribution may be legally applied. The probability that the value of the random variable assumes a value k is given by:

$$p(X = k) = \lim_{n\to\infty} [n!/(k!(n - k)!)\cdot(\lambda/n)^k\cdot(1 - (\lambda/n)^{n-k}]$$

$$\begin{aligned}
&= (\lambda^k/k!)\cdot\lim_{n\to\infty} [(n^{-k})(n)(n-1) \ldots (n-k+1)\cdot \\
&\quad (1-(\lambda/n)^{n-k})] \\
&= (\lambda^k/k!)\cdot\lim_{n\to\infty} [(1)(1-1/n)(1-2/n) \ldots \\
&\quad (1-(k-1)/n)\cdot(1-(\lambda/n)^{n-k})] \\
&= (\lambda^k/k!)\cdot\lim_{n\to\infty} [1-(\lambda/n)^n] \\
&= (\lambda^k/k!)\exp(-\lambda) \qquad (3.12)
\end{aligned}$$

We introduce now a notation to distinguish between these two probability distributions as follows:

1. $B(n,p,k)$ is the probability that the random variable X will assume a value k according to the binomial distribution when the sample size is n and the proportion defective (or probability of a given state for one outcome in the sample) is p.
2. $P(\lambda,k)$ is the probability that the random variable X assumes the value k according to the Poisson distribution when the average number of occurrences of the event which X describes on a given continuum is λ. We note that

$$B(n,p,k) = nCk(p)^k(1-p)^{n-k} \qquad (3.13)$$

and

$$P(\lambda,k) = (\lambda^k/k!)\exp(-\lambda) \qquad (3.14)$$

Example 3.4

It is known that the average number of failures of a given system is approximately constant at a value of twice per year. What is the probability that in a year there will be no failures; and what is the probability of more than two failures in any single year?

We assume that the Poisson probability distribution applies, i.e. failures occur at random and that an approximately constant average is definable for a given continuum. In our case the continuum is one year and the average has a value of 2, i.e. $\lambda = 2$. We wish to determine the probability for the random variable to assume first a value of zero and, secondly, a value greater than 2:

1. $P(\lambda, k)$ for $\lambda = 2$, $k = 0$, gives
   $$P(2, 0) = \exp(-2)2^0/0! = \exp(-2) = 0.1353$$
2. $P(\lambda, k)$ for $\lambda = 2$, $k > 2$, gives
   $$\begin{aligned} P(2,k > 2) &= 1 - [p(2,0) + p(2,1) + p(2,2)] \\ &= 1 - [\exp(-2) + 2\exp(-2) + 2\exp(-2)] \end{aligned}$$

$$= 1 - \exp(-2)$$
$$= 0.3233$$

The reader should now attempt to solve the problem of Example 3.3 using, on this occasion, the Poisson probability distribution, and compare his result with the earlier one. It should be noted from this that the Poisson is a good approximation to the binomial when n is large (say greater than 100) and p is small (say less than 0.01).

3.3.3 Negative exponential distribution – continuous random variable

Although the binomial and Poisson distribution formulae are well known and very approximately model a great many physical processes, they are by no means the only formulae available. Indeed a large number of probability distributions have been developed to model with greater accuracy a wide range of physical processes encountered in the decision-making and design considerations relating to large-scale systems problems. These two formative distributions are of the same type, i.e. discrete random variable. A different type is concerned with the continuous random variable, an example of which is the interval of time between arrivals or the interval of space between defects; in other words, the intervals on a continuum between successive or adjacent events which are randomly distributed on that continuum. We shall now show that where the distribution of an event on a continuum is given by the Poisson formulae, then the continuous random variable probability distribution for intervals between succeeding or adjacent events on that continuum is given by the negative exponential probability function.

Our first point to note is that inasmuch as a continuous variable may assume any of an infinity of values, the probability that it assumes any one of these must be infinitessimal. We must state then that probability values can only be assigned to the random variable assuming a value within a range of values of the random variable and, furthermore, that the probability value tends to zero as the range tends to a single value. We then ask 'what the probability is that the random variable of interest will assume a value in the range tA to tB, assuming both of these to be amounts of lapsed time to the succeeding event on a continuum starting from the immediately preceding event.' Put another way, what is the probability that the inter-event time on a continuum, when the distribution of such events is given by Poisson, will fall within a range between tA and tB ($t\text{A} < t < t\text{B}$)? Let the probability be denoted by p_{AB}. Then the probability of the converse, that the inter-event time falls outside this interval, is $1 - p_{AB}$.

This probability term p_{AB} is the sum of two probabilities: $p1$ and $p2$ relating to two mutually exclusive outcomes. The first is the outcome that the inter-event time will be less than tA whose probability of occurrence is, say, $p1$; the second is the outcome that the inter-event time will be greater than tB, whose probability of occurrence is, say, $p2$. Let us now assume that the average for the continuum is λ, then for a subcontinuum of length tA, the average is λtA, and for a subcontinuum of length tB the average is λtB. Using Poisson, the probability of zero events in a time interval of length tB is

$$\exp(-\lambda tB)$$

and this is exactly $p2$. Using Poisson again, the probability of anything other than zero events in a time interval of length tA is

$$1 - \exp(-\lambda tA)$$

and this is exactly $p1$. Therefore we write

$$1 - p_{AB} = p1 + p2 = \exp(-\lambda tB) + 1 - \exp(-\lambda tA)$$

Therefore

$$p_{AB} = \exp(-\lambda tA) - \exp(-\lambda tB) = \exp(-\lambda tA)\cdot[1 - \exp(-\lambda(tB - tA))]$$

This is a very useful result since it defines a kind of probability distribution formula for the random variable of interest. However, it would be better to have a general formula with which this result and others of interest could be generated. What we seek is a probability density formula which can be used to tell us all the probability terms concerning the continuous random variable.

Let us assume that we are primarily interested in an infinitessimal range about the point of elapsed time t. Then let $tA = t$ and $tB = t + dt$. This gives p_{AB} as follows:

$$\begin{aligned} p_{AB} &= \exp(-\lambda t) - \exp(-\lambda(t + dt)) \\ &= \exp(-\lambda t)[1 - \exp(-\lambda dt)] \\ &= \exp(-\lambda t)\cdot[1 - (1 - \lambda dt + 0\ (dt^2))] \\ &= \lambda \exp(-\lambda t)dt \end{aligned}$$

The term p_{AB} is now an infinitessimal probability, as expected, involving the infinitessimal of dt. In graphical terms that part of the expression for p_{AB} which is $\lambda \exp(\lambda t)$ denotes a height which remains approximately constant over a length t. Multiplying the two will give an approximated value for the area of the rectangle denoting the infinitessimal probability value p_{AB}. The 'height' term is known as the probability density function. It is definitely not itself a probability but rather it defines a curve under which lie areas, between certain values of the variable t, denoting precisely the probability that the continuous random variable will assume a value within those limits. This variable t now assumes the identity of the continuous random variable, inter-event, time for events distributed on a continuum according to a Poisson distribution and the function $f(t)$ given by equation 3.15 is known as the negative exponential probability density function:

$$f(t) = \lambda \exp(-\lambda t) \qquad (3.15)$$

The reader should now confirm that the negative exponential distribution formula is legal; that its expected value is $1/\lambda$; and that its variance is $1/\lambda^2$. The formulae that are required to be proved are:

$$\int \lambda \exp(-\lambda t) dt = 1 \qquad (3.16)$$

$$\int \lambda t \exp(-\lambda t) dt = 1/\lambda \qquad (3.17)$$

$$\int lt^2 \exp(-\lambda t) dt = 2/\lambda^2 \qquad (3.18)$$

all integrals taken from 0 to ∞.

3.3.4 Other distribution laws

The negative exponential formula brings our total of familiar probability distribution or density functions to three. Many more are available for the analysis of systems in which uncertainty is a major characteristic. The following are just a few examples of those available and are given for the reader's benefit in order to provide a little more appreciation of the sophistication that this probability theoretical approach has to offer the systems engineer in his studies. They are merely presented here for advance information and are mentioned again in the context of reliability in the next chapter. They are as follows:

1. Weibull

$$f(x) = \beta\, \lambda^{-\beta}\, x^{\beta-1} \exp[-(x/\lambda)^{\beta}] \qquad (3.19)$$

$$x > 0,\ b > 0,\ \lambda > 0$$

2. Lognormal

$$f(x) = 1/(x\sigma\sqrt{2\pi}) \exp[-(\log(x - \mu)^2/2\sigma^2] \qquad (3.20)$$

$$0 < x < \infty$$

3. Normal

$$N(\mu, \sigma) \text{ or } f(x) = 1/(\sigma\sqrt{2\pi}) \exp[-(x - \mu)/2\sigma] \tag{3.21}$$

$$-\infty < x < \infty$$

4. Standard–normal

$$N(0, 1) \text{ or } f(z) = 1/(\sqrt{2\pi}) \exp[-z^2/2] \tag{3.22}$$

$$-\infty < z < \infty$$

5. Gamma

$$f(x) = 1/(\lambda^{\beta}\Gamma(\beta))\, x^{\beta-1} \exp(-x/\lambda) \tag{3.23}$$

$x > 0$, where

$$\Gamma(\beta) = \int x^{\beta-1}\, dx, \qquad \text{i.e. } \Gamma(\beta + 1) = B\Gamma(\beta) \tag{3.24}$$

6. Special Erlangian

$$f(x) = x/\lambda^2 \exp(-x/\lambda) \tag{3.25}$$

$$x > 0$$

This is the gamma distribution with $\beta = 2$.

3.4 APPLICATIONS

To conclude this chapter we consider the application of probability theory to two imaginary situations. Notwithstanding the simplifications made, these examples offer scope for modeling processes which illustrate the competitive and traffic aspects in systems problems, respectively. Though the models themselves cannot be taken as serious examples of real-life situations, they give the reader an opportunity to apply probability analysis and permit therefore an evaluation of these processes.

3.4.1 Conflict strategies simulation

The first situation concerns the assignment of a destructive capability, a missile command force, to specified enemy targets with the objective of maximizing the kill ratio with respect to the parameters that describe the problem in an elementary force. Let us suppose that the number of weapons is n, the number of targets is m and, for convenience, n/m is an integer. The probability that a weapon will destroy a target on a given weapon/target encounter is denoted by p.

Two strategies are suggested. The first, X, is that weapons are randomly assigned to targets which can easily lead to an 'overkill' situation and to targets not being attacked at all. A more expensive strategy, Y, is where weapons are 'allocated' to targets and controlled to their respective assignments by specific equipment. The real problem is now to evaluate and compare the different strategies in the general context of the parameters, n, m and p.

In the case of the random weapon assignment, the following will hold:

1. Each target has a chance of being hit by a given weapon of $1/m$.
2. The probability of a kill on an encounter for each target is therefore p/m (probabilities of independent events).
3. The probability that a target will survive an encounter is therefore $1 - p/m$ (probabilities of mutually exclusive events).
4. The target may be hit by any of n weapons so that the probability that it survives n weapons is $(1 - p/m)^n$ (independent events again).
5. The probability that a target will not survive n attacks is therefore $1 - (1 - p/m)^n$ (mutually exclusive events).
6. This represents the likelihood of any single target being destroyed.

Therefore the expectation of number of targets destroyed must be

$$\mathrm{Ex} = m \cdot [1 - (1 - p/m)^n] \tag{3.26}$$

For the chosen values of n and m, Ex may be plotted against p so that the case for random assignment can be studied (simulated) for any desired likelihood of kill for a weapon/target encounter, which is itself a performance factor of the weapon used.

The alternative case, controlled assignment, is modeled as follows:

1. Each target is assigned n/m weapons.
2. Probability of one weapon destroying the target is p and therefore the probability of a target surviving one weapon is $1 - p$.
3. Probability of a target surviving all its weapons is $(1 - p)^{n/m}$.
4. Probability of target not doing so is $1 - (1 - p)^{n/m}$.
5. The expectation of targets destroyed is therefore

$$\mathrm{Ey} = m \cdot [1 - ((1 - p)^{n/m})] \tag{3.27}$$

Figure 3.3 shows the graphs of expectation per target for both strategies plotted against p on the interval 0, 1, for $n = 12$, $m = 3$. This might be called a simulation of the model for these parameter values since it reveals a comparative evaluation of the two strategies. For example, suppose a value of p to be $p1$ as shown. With strategy X the effectiveness has a value of A. To move to the more effective strategy Y, keeping p constant, would increase the effectiveness to B. The same effectiveness could be achieved by keeping to strategy X but by increasing the value of p to $p2$. The analyst is in a position now to evaluate, quantitatively, the relative merits of the two strategies in relation to purchasing a more effective weapon. Further complications could easily be catered for without making the model unnecessarily complex. The application of

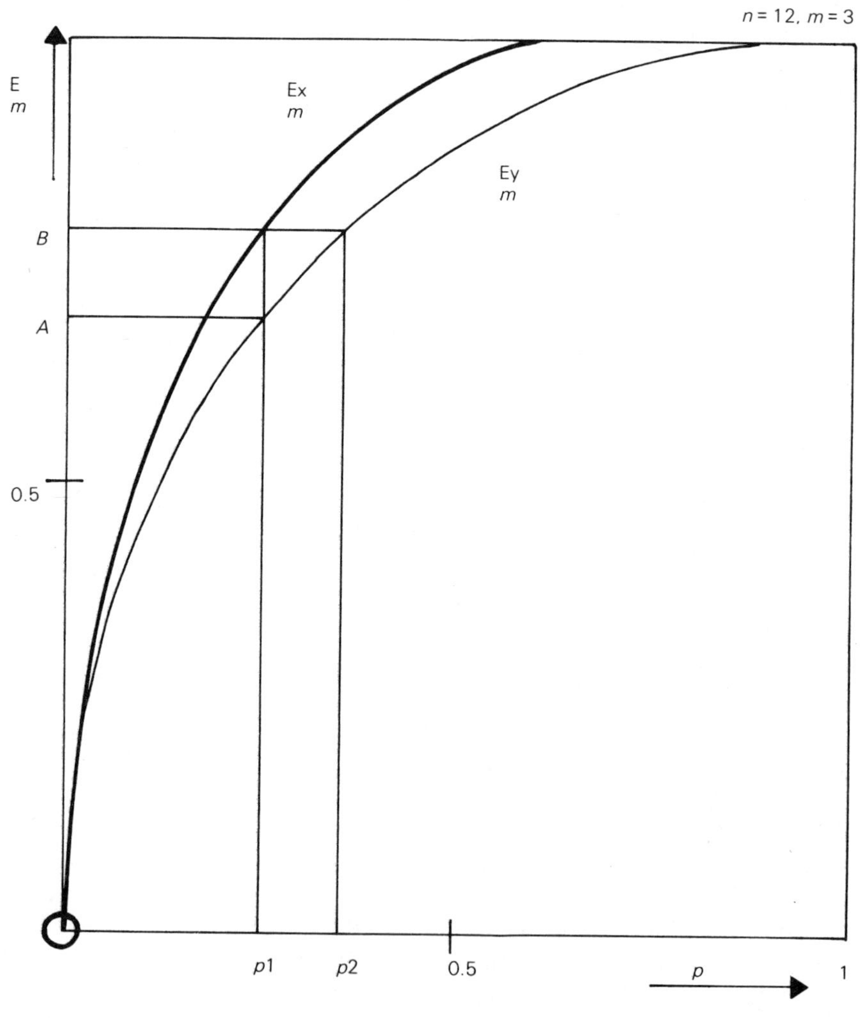

Figure 3.3

probability theory and the simulation of even an elementary model shows how uncertainties in a situation can be accommodated and approached in an analytical manner.

3.4.2 Waiting line simulation

We turn now to a second physical process involving uncertainty, but this time concerned with a waiting-line situation. A continuous supply of widgets awaits processing on machine A. Depending on a number of factors, processing time on A can be 5, 10 or 15 minutes. Having been serviced on A, widgets proceed to machine B where processing time takes either 5 or 10 minutes. A queue facility is available between the two machines in the event that B is occupied when widgets arrive there. Both machines can service only one widget at a time but they may do so simultaneously. The problem is to devise a model that will enable the calculation of idle time of machine B, i.e. the time that it is available to process widgets when there is none to process; the waiting time spent on average by a widget; and the average time spent by a widget in the system.

In Chapter 5 the analysis of queues is discussed showing how it is possible to go to great lengths with analytical rigid models before they become unmanageable in relation to the sophistication of the systems they describe. But the opportunity is taken here to describe a simple and convenient means of obtaining the answers required for very little analytical effort. It employs the Monte Carlo technique in simple form to provide a numerical probability model.

The Monte Carlo technique is a means of simulating physical (actual) stochastic events, governed by known probability distributions, by random number generation and selection routines. Suppose the probability distribution for t_A is:

$$p(t_A = 5) = 0.25, p(t_A = 10) = 0.5, p(t_A = 15) = 0.25$$

and that for t_B is:

$$p(t_B = 5) = 0.25, p(t_B = 10) = 0.75$$

By drawing random numbers in the range 0–99 the selection of any number is equally likely and therefore has event probability 0.01. Supposing therefore the number drawn was in the range 0–24, the event probability here is 0.25 which is the same as for $t_A = 5$ (and equals 15 for that matter). A scheme that decided the value for t_A on the basis of the drawing of a random number (RN) as shown below would be a simulation for the processing time on machine A:

$0 \leqslant RN \leqslant 24 \qquad t_A = 5\ (p = 0.25)$
$25 \leqslant RN \leqslant 74 \qquad t_A = 10\ (p = 0.5)$
$75 \leqslant RN \leqslant 99 \qquad t_A = 15\ (p = 0.25)$

A similar scheme for B would be

$0 \leqslant RN \leqslant 24 \qquad t_B = 5\ (p = 0.25)$
$55 \leqslant RN \leqslant 99 \qquad t_B = 10\ (p = 0.75)$

The simulation of the entire process will consist of these two schemes, requiring random numbers to be drawn successively for times on A and B and the orderly calculation of the following:

T_A, start time on A
T_F, finish time on A
T_B, start time on B
T_E, finish time on B (exit from system).

These values can be calculated for each widget and are therefore more properly represented as $T_{A(k)}$, $T_{E(k)}$, etc. The simulation is now described by the logical relations of these identifiers:

$T_{A(k)} = T_{F(k-1)}$; this widget starts on A immediately the last finishes (no idle time for machine A).
$T_{F(k)} = T_A + t_{A(k)}$; this widget finished on A $t_{A(k)}$ later.
$T_{B(k)} = \max\,[T_{F(k)}, T_{E(k-1)}]$; this widget starts on B as soon as it arrives or when the previous one finished on B, whichever is later. Note that if $T_{F(k)} > T_{E(k-1)}$ widget waits.
$T_{F(k)} < T_{E(k-1)}$ is idle
$T_{F(k)} = T_{E(k-1)}$ neither case.
$T_{E(k)} = T_{B(k)} + t_{B(k)}$; this widget leaves the system when it is finished on B.
$T_{s(k)} = T_{E(k)} - T_{A(k)}$; time spent by this widget in the system.
$T_{w(k)} = [T_{F(k)} - T_{E(k-1)}]\cdot\delta$; widget waiting time, where $\delta = 0$ if $[T_{F(k)} - T_{E(k-1)}]$ is positive and 1 otherwise.
$T_{i(k)} = [T_{E(k-1)} - T_{F(k)}]\cdot\delta$; idle time for B on this widget.

Table 3.1 shows a typical distribution of events for this simulation. From this table, the necessary average values of the terms of interest can easily be calculated: average idle time = 1.875 min.; average waiting time = 1.875 min.; average time spent in system = 20.125 min. Table 3.2 shows the machine characteristics.

Table 3.1 Digital simulation of production process

T_A	t_A	T_F	T_B	t_B	T_E	T_S	T_w	T_i
0	10	10	10	5	15	15	0	10
10	10	20	20	10	30	20	0	5
20	10	30	30	5	35	15	0	0
30	10	40	40	5	45	15	0	5
40	15	55	55	10	65	25	0	10
55	10	65	65	10	75	20	0	0
65	5	70	75	5	80	15	5	0
70	10	80	80	5	85	15	0	0
80	10	90	90	10	100	20	0	5
90	5	95	100	10	110	20	5	0
95	5	100	110	10	120	25	10	0
100	10	110	120	10	130	30	10	0
110	10	120	130	5	135	25	10	0
120	10	130	135	10	145	25	5	0
130	10	140	145	5	150	20	5	0
140	10	150	150	10	160	20	0	0
150	15	165	165	10	175	25	0	5
165	10	175	175	10	185	20	0	0
175	5	180	185	10	195	20	5	0
180	15	195	195	5	200	20	0	0
195	15	210	210	10	220	25	0	10
210	10	220	220	10	230	20	0	10
220	20	230	230	5	235	15	0	0
230	10	240	240	10	250	20	0	5
240	10	250	250	10	260	20	0	0
250	5	255	260	10	270	20	5	0
255	10	265	270	5	275	20	5	0
265	15	280	280	5	285	20	0	5
280	5	285	285	5	290	10	0	0
285	10	295	295	10	305	20	0	5
295	10	305	305	10	315	20	0	0
305	10	315	315	10	325	20	0	0
315	10	325	325	5	330	15	0	0
325	10	335	335	10	345	20	0	5
335	15	350	350	10	360	25	0	5
350	10	360	360	10	370	20	0	0
360	5	365	370	10	380	20	5	0
365	10	375	380	10	390	25	5	0
375	15	390	390	10	400	25	0	0
390	1	400	400	5	405	15	0	0

Table 3.2 Machine characteristics

Production time details Machine A		Machine B	
t_A	$p(t_A)$	t_B	$p(t_B)$
5	0.25	5	0.25
10	0.50	10	0.75
15	0.25		

3.5 CONCLUSIONS

This chapter has introduced the basic notions which have contributed to the development of probability theory. These are the sample space, the probability axiom, the probability law and the random variable. The application of probability theory to actual physical processes has called for extensive developments in probability distribution formulae and, combined with the subject of mathematical statistics, the systems engineer is now provided with an enormously powerful tool for the modeling and simulation of those processes. The challenge of uncertainty, to be found everywhere in systems engineering studies, is well matched by those rigorous topics whose foundation is probability theory. Two prominent examples of these topics, reliability and queues, are the subject of the following chapters.

3.6 EXERCISES

1. Packets of food are filled automatically and the proportion of underweight packets in a very large batch is p. A sample size n is selected randomly in the batch and the probability that the sample contains exactly r defective packets ($r = 0, 1, 2, \ldots, n$) follows a certain probability distribution. Name this distribution and write down the probability that the sample contains exactly r defective packets. For one particular process it has been found that 2 percent of the packets are underweight.

An inspector takes a random sample of 10 packets. Calculate:

(a) The expected number of packets in the sample which are underweight.
(b) The probability that none of the packets in the sample is underweight.
(c) The probability that more than one of the packets in the sample is underweight.

2. It has been found in the past that 4 percent of the screws produced in a certain factory are defective. A sample of 10 is drawn randomly from each hour's production and the number of defectives noted. In what fraction of these hourly samples would there be at least two defectives? What doubts would you have if a particular sample contained six defectives?

3. Of a certain type of car, 1 percent have a defective tail light. How many cars must be inspected to have more than an even chance of finding a defective tail light?

4. An electronic component is mass-produced and then tested unit by unit on an automatic testing machine which classifies the unit as 'good' or 'defective'. But there is a probability of 0.1 that the machine will misclassify the unit, so that each component is in fact tested five times and regarded as good if so classified three or more times. What now is the probability of a misclassification?

5. A recent court decision supported a pilot's decision to continue to his destination with a four-engine jet aircraft after an engine failure at midrange on a two-hour flight, rather than land at the midway point. His argument was that the aircraft will fly on two engines and that the probability of two additional failures was quite small. If the one-hour reliability of a single engine is 0.999, do you agree with the pilot's decision? If the aircraft was a three-engine jet that will fly on one engine, would you agree? Comment on the comparative reliability of three- and four-engine jets.

6. The new 'Cheetah' sports car has an idle loping problem as about 10 percent of the Cheetahs have an unstable fluctuating engine speed when they are idling. An engineering 'fix' is put in a production pilot lot of a 100 cars. (a) If the fix has no effect on the problem, how many cars would you expect to have the fault? (b) If only 2 of the pilot lot have the idling fault and the other 98 cars are not defective, would you conclude that the fix has a significant effect? (Hint: show that the probability of 2 or less defectives in a sample of 100, given that the fix has no effect, is very small.)

7. The average number of calls that a hospital receives for an ambulance during any half-hour period is 0.20. Considering a reasonable cost per ambulance and crew and presuming that any ambulance will return to the hospital in half an hour, how many ambulances would you recommend for this hospital? Comment on the idea of ambulance pools which are shared by several hospitals.

8. A manned interplanetary space vehicle has four engines each with reliability 0.99. Each engine has a failure detection system which may itself fail. If the engine does fail there is a conditional probability of 0.02 that a success will be signalled. If an engine fails and is not detected the result is catastrophic. However, the mission can be completed with three engines if one engine fails and is detected. What is the probability of mission success? If there is no abort system or escape system, which action would you recommend if two failures are signaled?

9. It has been found in the past that 1 percent of electronic components produced in a certain factory are defective. A sample of size 100 is drawn from each day's production. What is the probability of getting no defective components by using: (a) the binomial distribution, and (b) the Poisson distribution approximation?

10. A construction company has a large fleet of bulldozers. The average number inoperative at a morning inspection owing to breakdowns is two. Two standby bulldozers are available. If a bulldozer can always be mended within 24 hours of the morning inspection, find the probability that at any one inspection: (a) no standby bulldozers will be required; and (b) the number of standby bulldozers will be insufficient.

3.7 FURTHER READING

Beaumont, G.P. (1983) *Introductory Applied Probability*, Ellis Horwood.

Beaumont, G.P. (1986) *Probability and Random Variables*, Ellis Horwood.

Feller, W. (1968) *An Introduction to Probability Theory and its Applications*, Wiley.

Miller, I. and Freund, J.E. (1985) *Probability and Statistics for Engineers*, Prentice-Hall.

Ross, S.M. (1984) *A First Course in Probability*, Macmillan.

Rowntree, D. (1984) *Probability*, Edward Arnold.

Rozanov, Y.A. (1977) *Probability Theory – a Concise Course*, Dover.

4 Systems reliability

4.1 RELIABILITY AS A VIRTUE

Reliability in people is undoubtedly a desirable quality and greatly to be admired in any individual. Few would disparage the reliable person or decry that person's worth in industry, commerce or society generally. In this respect reliability is synonymous with trustworthiness, dependability and consistency, terms which underline the fact that the characteristic of reliability is highly valued. There are, of course, degrees of reliability and on a comparative basis an adjudication of which individual is more reliable than another is easily made. On the other hand, it is difficult to define the characteristic precisely or to measure its value or worth exactly. When one considers that the characteristic of reliability is usually used to describe some function or task, it becomes slightly easier to estimate the measure of an individual's reliability in relation to that function or task, e.g. record of attendance at meetings.

It is not difficult to see how or why the general concept of reliability has been applied to the man-made world. Just as people can let others down, so also functional objects intended for a particular purpose can fail to perform in the manner desired. The unreliability of the world's artefacts account for large amounts of wasted time and money and the endangerment of human life. As the consequences of this type of unreliable behavior become more serious, so the interest in reliability and the desire for more reliable products increases. The degree of interest and the standard of reliability are obviously closely coupled to the consequences of unreliable behavior. Moreover to raise the standard of reliability will, in general, involve increased expenditure. As with all practical problems, a balance must be struck between an acceptable degree of reliability in a product's performance and the amount of money which can be spent in achieving such. In other words, at some point one settles for the advantages to accrue from a level of reliability and does not invest further in adding to these advantages.

The question is therefore posed about every item where reliability is of interest – 'is it reliable enough?' This inevitably leads to the conclusion that to achieve an economic balance, reliability itself must be defined and hence used in a way that it becomes a measurable quantity. A review of the way in which the interest in reliability engineering has developed reveals four important conclusions:

1. The unreliability of any functional item or technological project can cause financial loss, human frustration, dissatisfaction and death.
2. The degree of required or acceptable reliability depends on the consequences of failure.
3. A formal scheme for the evaluation of system reliability is essential if the risks and consequences of failure are to be balanced against the benefits of achieving higher degrees of reliability.
4. The sophistication of reliability evaluation must increase to take account of those technological projects in which the risks of failure are considerably greater, i.e. high reliability systems.

Examples of high risk situations where there is either a large amount of capital invested, or a danger to human life, or both, would include the following:

1. Aircraft systems (in particular the latest types of instrument landing systems).
2. Nuclear power plants: a failure could mean either that life is endangered or that there is a significant loss of revenue, even if the plant is unavailable for a short duration only.
3. Chemical plants: these have become more complex and capital intensive in the search to produce more of the less easily obtained chemicals and to improve efficiency of plant operation. Accordingly the greater energy requirement of these modern plants gives rise to the possi-

bility of greater plant damage under fault conditions. Damage can mean direct and indirect economic loss, the latter due to the loss of future markets, and could involve risk to human life both within the plant and outside as a consequence of the emission of poisonous materials in some processes.

4. Electrical supply systems: since electricity is an indispensable commodity in modern society, both its cost and availability are of prime importance. Costs and associated reliability relate to all aspects of the supply system, i.e. generation, transmission and distribution, by means of which the customer finally gets his service. Loss of supply can be very much more than just in economic embarrassment – in some situations it can also put human life at risk.

As the particular situations of these systems are discussed in some detail, certain basic design principles emerge. Chiefly, these are the use of diversity, redundancy and designing with a preferred mode of failure in mind (such as always endeavoring to produce a fail-safe condition).

Obviously it becomes necessary to compare the various approaches in any given design situation on the grounds of reliability. Some designers may investigate various lines of action to improve reliability, others may wish to show whether a particular system is adequately reliable. For this to be approached in a quantitative and logical fashion, the definition of reliability based on the idea of probability needs to be developed. One such definition, couched in probabilistic terms, is offered:

> Reliability is defined as that characteristic of an item expressed by the probability that it will perform its required function in the desired manner under all relevant conditions and on the occasions or during the time intervals when it is required so to perform.

An important aspect of reliability, contained in the above definition, is to see how probabilistic concepts may be applied to events occurring in the time domain. We shall see later how this is developed. First, however, we take a simplistic view which allows us to consider another important aspect of reliability performance: availability.

4.2 A FREQUENCY INTERPRETATION: AVAILABILITY

If N failures of an item occur in some total time of interest T, then, to keep in line with the definition of reliability given above, it is necessary to know the probability of failure as some function of N and the time T.

The total time T may be considered as being made up of n equal subintervals of length $\mathrm{d}t$. The time increment $\mathrm{d}t$ needs to be chosen so that any increment contains either no failures or not more than one failure. Each time increment may be looked on as being either a successful or a failed event therefore. Using the frequency concept of probability, the value P_f, the probability of any time increment $\mathrm{d}t$ being classified as a failure event, is:

$$P_f = \lim_{n\to\infty} (N/n) \tag{4.1}$$

Since both N and T have been taken to be finite, the $P_f \to 0$ as $\delta t \to 0$, which is a result one must expect, i.e. there is zero probability of a failure occurring in an infinitessimal period of time. However, there is obviously some finite probability of failure related to the total time T and this may be deduced by considering the limiting combination of the probabilities of failure for each time increment δt.

The combined probability $P(T)$ of at least one failure occurring in T is made up of the probability of a failure occurring in any one or more of the increments of time t. Hence:

$$P_s = \lim_{n\to\infty} ((n - N)/n) \tag{4.2}$$

where $P_f + P_s = 1$

$$1 - P(T) = \lim_{n\to\infty} (1 - N/n)^n$$
$$P(T) = 1 - \lim_{n\to\infty} (1 - N/n)^n = 1 - \exp(-N) \tag{4.3}$$

provided that the value of P_f and hence N can be taken as being constant and time independent relative to the position of any time increment $\mathrm{d}t$ in the complete time range T. Since there is no systematic spreading or bunching of failures in the time domain, then it is possible to define a mean failure-rate parameter, Θ, such that:

$$\Theta = N/T \tag{4.4}$$

where

$$p(t) = 1 - \exp(-\Theta t) \tag{4.5}$$

where t is now any time interval of interest over which the value of u may be taken as being time independent or constant and $p(t)$ is the probability of at least one failure occurring in this time interval, it being assumed that no failure was known to exist at time zero. Note that when $\Theta t \ll 1$ then $p(t) \approx \Theta t$.

Example 4.1

A digital control system consists of ten printed circuit boards. The past records show that the average number of times each board breaks down under normal conditions is once in every two years, and this value remains constant over long periods of time. What are likely to be the average number of calls on the breakdown service in a year? What is the probability of at least one breakdown in any three-monthly period?

A mean failure-rate Θ, can be assigned to each board and taking the figures of one fault in a two-year period:

$$\Theta = 1/2 = 0.5 \text{ breakdowns/year}$$

For a system of ten boards, the overall failure-rate is:

$$\Phi = 10{\cdot}\Theta = 5 \text{ breakdowns/year}$$

The probability of at least one breakdown in any three-monthly period may be calculated as follows:

$$\Phi = 5 \text{ breakdowns/year}$$
$$t = 1/4 \text{ year (3 months)}$$
$$p(t) = 1 - \exp[-(5 \times 14)] = 0.714$$

giving a 71.4 percent chance of at least one breakdown in any three-monthly period. The concept of availability becomes important when failures lead to some finite length of time associated with repairing, restoring or replacing the failed item.

In the simplest case it may be assumed that, when an item fails, it is out of action for some fixed length of time $\mathrm{T_r}$ which represents the repair time or its equivalent. If it is taken that N failures occur in a total time T, then the total 'downtime' associated with the interval is:

$$T_\mathrm{d} = N\mathrm{T_r} \tag{4.6}$$

and similarly, the total available time, T_a, is given by:

$$T_\mathrm{a} = T - N\mathrm{T_r} \tag{4.7}$$

A quantity, mean fractional dead time (or downtime), D, is defined which represents the mean proportion of dead time over the total time of interest:

$$D = \text{mean time in the failed state/total time}$$

In the case of the simple failure and repair model, it is seen that:

$$D = NT_r/T \tag{4.8}$$

and a limiting case of this mean fractional dead time may be considered where:

$$D = \lim_{T\to\infty} (NT_r/T) \tag{4.9}$$

where D has characteristics similar to those of probability as suggested by the frequency probability model.

Mean availability, A, may be defined:

$$A = \text{mean time in the working state/total time}$$

and hence, again for our simple repair model:

$$A = \lim_{T\to\infty} ((T - NT_r)/T) \tag{4.10}$$

i.e. $A + D = 1$.

Provided that the ratio N/T can be taken as being time-invariant, then the number of failures which occur may be expressed in terms of a mean failure rate. In the case of our simple repair model, the failure-rate characteristic Θ refers to the actual working time T_a and not to T, hence:

$$N = \Theta T_a \tag{4.11}$$

If we substitute for T_a using equation (4.7):

$$N = \Theta[T - NT_r]$$

or

$$N = \Theta T/(1 + \Theta T_r) \tag{4.12}$$

and then from (4.9):

$$D = \Theta T_r/(1 + \Theta T_r) \tag{4.13}$$

or

$$D \approx \Theta T_r \textbf{ if } \Theta T_r \ll 1$$

Similarly

$$A = 1/(1 + \Theta T_r) \tag{4.14}$$

These are simple expressions for mean fractional dead time and mean availability.

For those cases where both the failure-rate characteristic and the repair process are themselves functions of time and for where different methods of detecting failures and carrying out repairs are taken into account, the expressions become more elaborate and the reliability evaluation more sophisticated. For such treatment the reader is advised to consult the further reading list.

Example 4.2

An electrical supply system is subject to failure which causes loss of supply to the consumer. The mean time between such failures is known to be 398 h and the mean time to repair the failures and restore the supply is known to be 2 h. What is the average value of the availability of the supply to the consumer over a long period of time?

If a mean time between failures of 398 h, which is time-independent, the corresponding value of the failure-rate Θ is given by:

$$\Theta = 1/398 \text{ failures/h}$$

and for a repair time T_r of 2 h using equation (4.14) we get:

$$A = 1/(1 + (2/398)) = 398/400 = 99.5\%$$

Example 4.3

A fog lamp on a car, which is not normally switched on, is subject to failures of the bulb and wiring which would cause the fog lamp to fail to light when required to do so. Such failures, which occur at a mean rate of one failure every 38 weeks, are not directly revealed but stay, on average, in existence for a mean period of 2 weeks until they are rectified. If, over some period of time, the chance of the driver of the car running into fog is 1 in 100, what is the chance that the fog light will not come on at the time it is required?

First we must calculate the probability of the fog lamp being in the failed state, which is given by the mean fractional dead time D, where

$$\Theta = 1/38 \text{ failures/week}$$
$$T_r = 2 \text{ weeks}$$

then

$$D = \frac{(1/38)2}{1 + (2/38)} = 2/40 = 0.05$$

The probability of fog P_f is 0.01. The probability of the simultaneous occurrence of the independent events 'fog' and 'lamp in failed state' is the product of the event probabilities and gives P, the probability that the fog lamp will not come on when it is required. Thus

$$P = 0.01 \times 0.05 = 0.0005$$

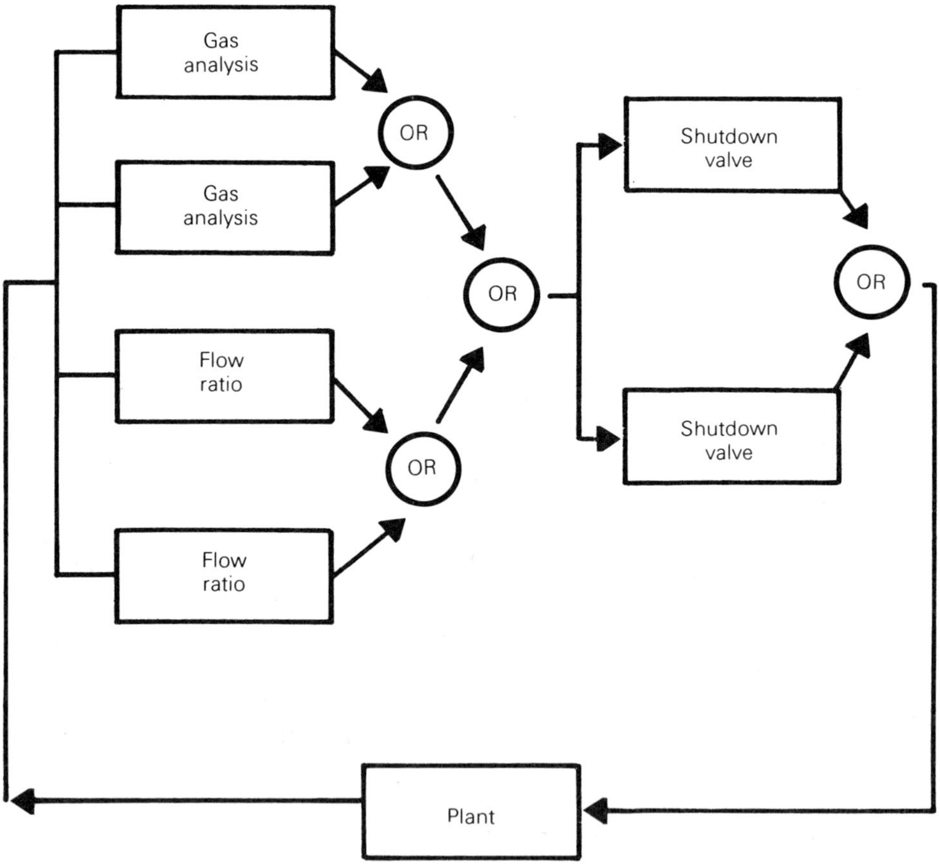

Figure 4.1 Automatic shutdown system.

Example 4.4

A chemical plant with a certain throughput of process gases is subject to an explosion hazard if the gas concentrations deviate outside certain specified limits. The plant is protected by an automatic shutdown system which measures both flow ratios and gas concentrations, and operates shutdown valves if either of these measurements moves outside the required limits. A block schematic of the shutdown system showing the logical flow and combination of information is shown in Figure 4.1.

The logical OR boxes mean that the correct shutdown information is passed on if any of the inputs have a signal requiring shutdown. The gas analysis measurement, the flow ratio measurement and the shutdown valves are all duplicated so that, in each case, both items must fail before that particular function is invalidated. For simplicity, it is assumed that the significant failures that can occur in each block and prevent it operating can be represented by some mean failure-rate u, and that such failures are independent and not directly detected but have associated with them some mean time in the failed state, T_r. If, over the operating life of the plant there is a 1 in 10 chance that a plant condition will arise demanding protection by the automatic shutdown system, what is the corresponding chance that an explosion hazard will occur? Typical values of u and T are as shown in Table 4.1

Table 4.1

Functional block	u (faults/year)	T_r(years)
Gas analysis	3.0	0.04
Flow ratio	2.0	0.04
Shutdown valves	0.2	0.05

We need to know the probability P of the simultaneous occurrence of the two independent events: a demand on the shutdown system ($P_d = 0.1$) and the shutdown system being in the failed state at the time of the demand (probability denoted by D_s), where:

$$P = P_d \times D_s$$

The quantity D_s is made up of the mean fractional dead times of the various functional parts of the system. For a single gas analysis unit the value D is approximately

$$D = \Theta T_r = 3.0 \times 0.04 = 0.12$$

For the complete gas analysis section, the combined probability of it being in the failed state, D_g, is given by:

$$D_g = D^2 = 1.44 \times 10^{-2}$$

Similarly for the flow ratio section, D_f is given by:

$$D_f = (2 \times 0.04)^2 = 6.4 \times 10^{-3}$$

The values of D_g and D_f are then combined in another OR function to make up the complete measurement section, for which the mean fractional dead time, D_m, is given by:

$$\begin{aligned} D_m &= D_g \times D_f \\ &= 9.216 \times 10^{-5} \end{aligned}$$

In the case of the control section, the mean fractional dead time, D_c, is given by the product of the mean fractional dead times for each shutdown valve, i.e.:

$$D_c = (0.2 \times 0.05)^2 = 10^{-4}$$

A failure of either the measurement section or the control section will inhibit the operation of the automatic shutdown system. Therefore the probability of either or both being in the failed state is given by:

$$\begin{aligned} D_s &= D_m + D_c - D_m \cdot D_c \\ &= 1.9 \times 10^{-4} \end{aligned}$$

Therefore

$$\begin{aligned} P &= 0.1 \times 1.9 \times 10^{-4} \\ &= 1.9 \times 10^{-5} \end{aligned}$$

There is about 1 chance in 50,000 of an explosion hazard during the plant life.

4.3 FAILURE DENSITY FUNCTIONS

If we return now to our original definition we are reminded of how reliability is expressed as a probability value and that this value is related to a certain period of time during which a failure or non-failure is of interest. Clearly we are interested in knowing how to obtain this probability value for any given object and would be further interested to

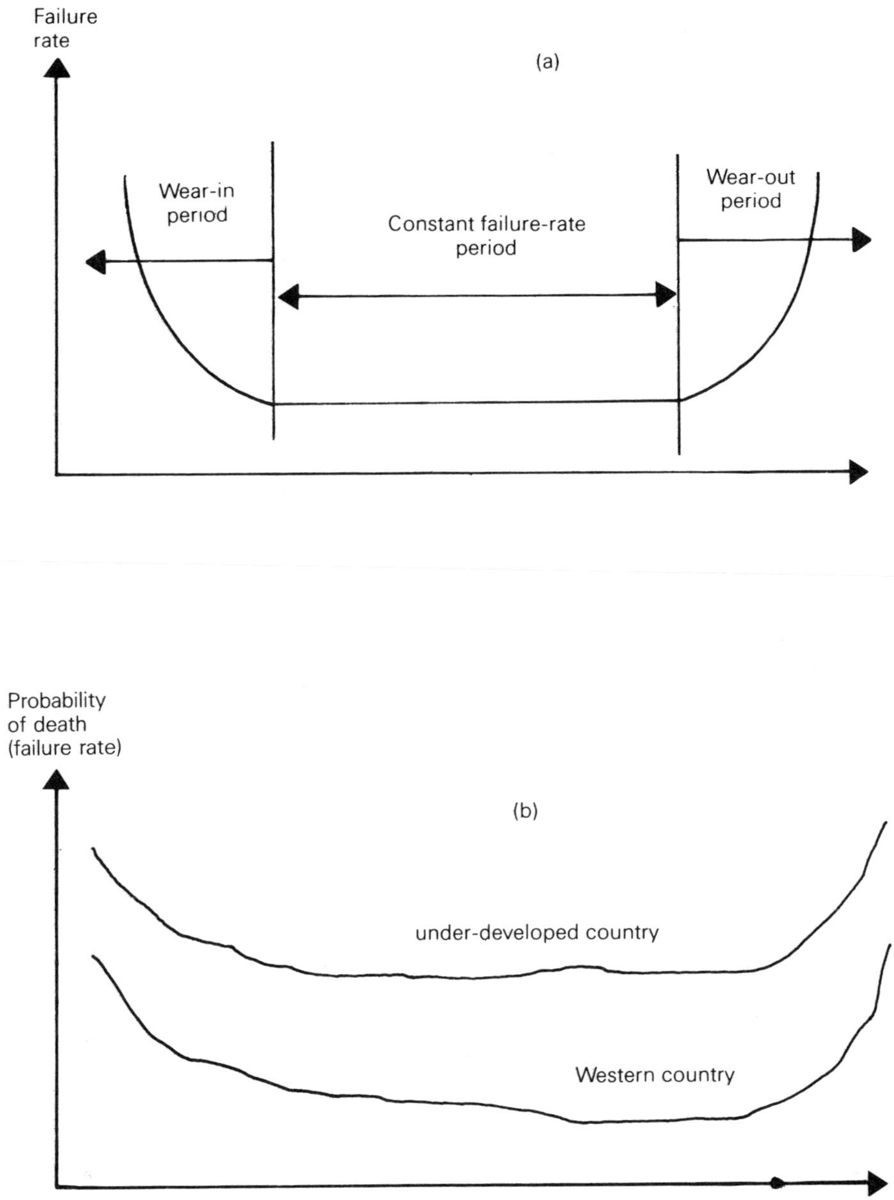

Figure 4.2 Failure-rate characteristics: (a) inanimate object; (b) humans.

know what relationship holds between this value and the extent of the period of time in question. An important concept in systems reliability which is a useful aid to discovering these facts of interest is that of failure rate. The mean failure-rate parameter Θ was introduced in equation (4.4) and this led to a definition of reliability as a probability value which decays exponentially with increase in the time for which the object of interest is expected to survive. In effect this assumes an approximately constant failure rate. It is normally assumed for any given object that failure rate will initially be high and will fall steadily through what is known as the wear-in phase. This will then be followed by a period in which the failure rate is maintained approximately constant, known as the 'useful life' of the object. Finally the failure rate increases once again in the wear-out phase. This characteristic of failure-rate dependence on the age of an object is summarized in the familiar bath-tub curve shown in Figure 4.2. A comparison is afforded by the failure rate for humans, i.e. probability of death dependent on the age of persons. The failure-rate function which is a probability density function is a useful device for determining relevant probability values as we shall now see.

Since the failure time of an item may be any positive number, the distribution of failure times is continuous. Thus it can be described either by its probability density function $f(t)$, also known as the failure density function, or by its cumulative distribution function $F(t)$, so that

$$F(t) = \int f(t)\mathrm{d}t \tag{4.15}$$

representing the probability of a failure within time t from zero. We introduce the reliability function $R(t)$ which gives the probability that an item will survive until time t, this also being the probability of no failure within t from zero. Thus

$$R(t) = 1 - F(t) \tag{4.16}$$

Now the probability that an item will fail in the interval to $t + \Delta t$ is

$$f(t)\cdot\Delta t = \text{probability element}$$

For this to be so, two events must take place, namely survival to t, then failure. Survival to t is $R(t)$. Therefore the conditional probability that an item will fail given that it has survived t, is

$$(f(t)/R(t))\cdot\Delta t$$

and we write:

$$Z(t) = f(t)/R(t) \tag{4.17}$$

where $Z(t)$ is the conditional failure rate function, or hazard function.

$Z(t)$ is always larger than $f(t)$, because $R(t)$ is always less than 1. For example, the proportion of cars which fail between 15 and 16 years old is very small, because so few last 15 years. However, of those that do survive, the proportion which then fail is very high.

Now

$$f(t) = \mathrm{d}F/\mathrm{d}t \tag{4.18}$$

and since

$$R = 1 - F, f = -\mathrm{d}R/\mathrm{d}t$$

Therefore Z being equal to f/R:

$$f = \mathrm{d}R/R\mathrm{d}t$$

Therefore

$$Z\mathrm{d}t = -\mathrm{d}R/R$$

Conversely

$$R = \exp[-\textstyle\int Z\mathrm{d}t] \tag{4.19}$$

Suppose now a product is such that its age has no effect on its probability of failing, i.e. $Z(t)$ = constant, say λ, so that if it has survived to t, the probability that it will fail during the next Δt is $\lambda\Delta t$. Items are now failing at a constant rate according to a continuous probability distribution. Then

$$R(t) = \exp(-\lambda t) \tag{4.20}$$

Therefore

$$F = 1 - \exp(-\lambda t) \tag{4.21}$$

and

$$f = \lambda\exp(-\lambda t) \tag{4.22}$$

as for negative exponential distribution.

The exponential model states that chance alone dictates when a failure occurs and that the product's age does not affect the issue. For many products this is not true and if these are to be looked at, it is necessary to examine more elaborate distributions, e.g. Weibull, gamma and Erlangian.

Figure 4.3 shows for each of four probability density functions the graphs for the cumulative distribution function, the reliability function and the hazard function.

In conclusion of this section on the time-dependent nature of reliability,

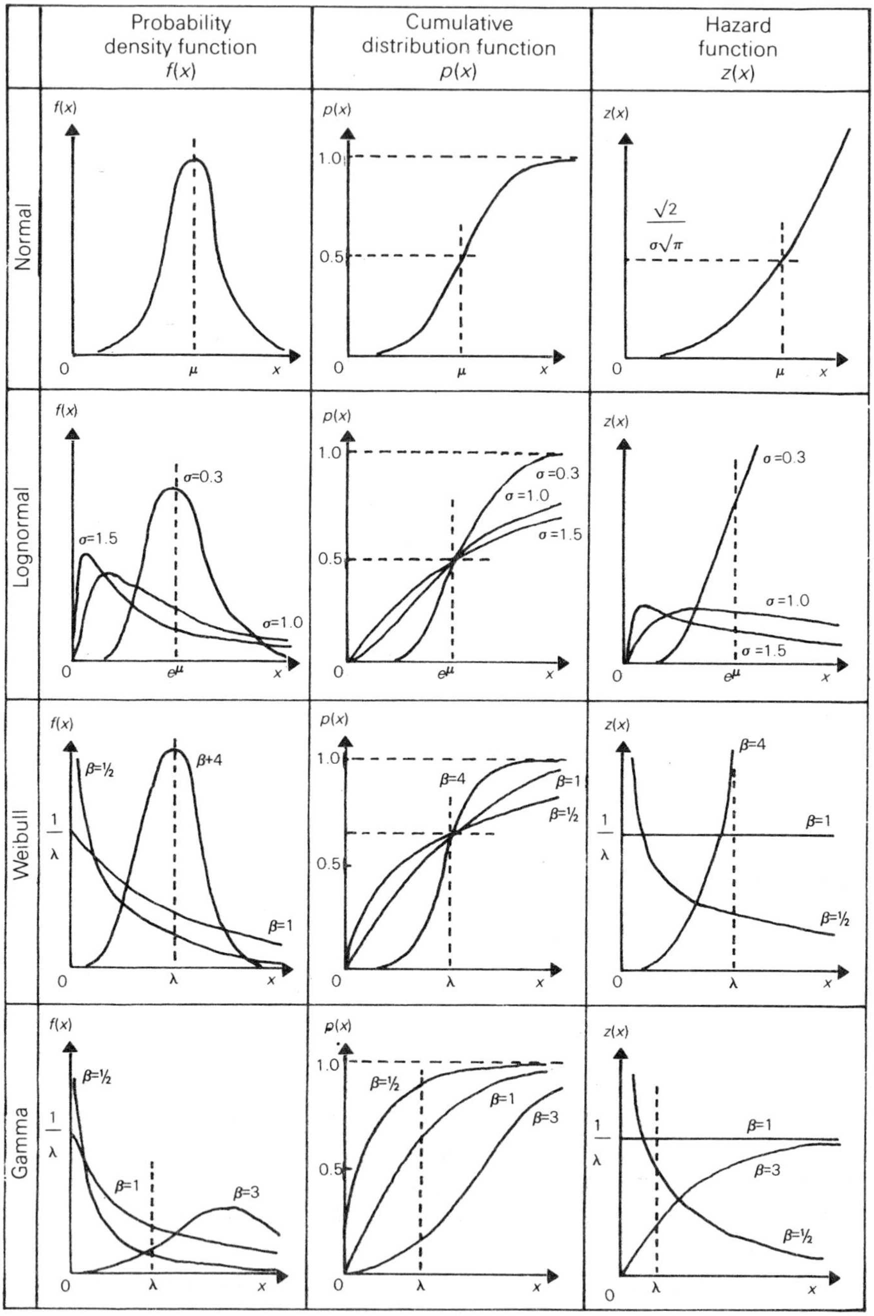

Figure 4.3 Probability distribution functions.

mention should be made of two important statistics, which are mean time between failures (MTBF) and mean time to failure (MTTF). The first applies to repairable items and means that if an item fails, say, five times over a period of use totalling 1,000 hours, the MTBF would be 1000/5 or 200 hours. The second applies to non-repairable items and means the average time an item may be expected to function before failure. This can be determined empirically by tests to destruction, for example. In these, a very large number of identical items are tested and data kept on the surviving items at given intervals of time.

Analytically these parameters are derived from a definition of the mean of the failure density function. Thus

$$\text{MTTF (or MTBF)} = \int t f(t) \mathrm{d}t$$

integrating from 0 to ∞.

In the case of the negative exponential distribution, the value of the MTTF or MTBF is simply the reciprocal of the average failure rate. This is so well known that the negative exponential failure density function is often written out in terms of its MTTF (τ), i.e. if $\tau = 1$, where λ is the approximately constant average failure rate, then

$$f(t) = \exp(-t/\tau)/\tau \tag{4.23}$$

Example 4.5

A resistor component in an electronic amplifier is subject to random and catastrophic failure. Over a large number of such components it has been deduced that the times to catastrophic failure are exponentially distributed with an MTTF of 10^7 hours. What are the respective probabilities that such a resistor will fail within the periods of: (a) 100 h, (b) 2,500 h, (c) 10,000 h and (d) 50,000 h?

If MTTF (τ) = 10^7 then failure rate $\lambda = 10^{-7}$ per hour. Failure density function for negative exponential is $\lambda\exp(-\lambda t)$ and reliability function is given by

$$R(t) = \exp(-\lambda t)$$

The failure function is:

$$F(t) = 1 - R(t) = 1 - \exp(-\lambda t) = 1 - \exp(-t/\tau)$$

If $t \ll \tau$, then $F(t) \approx t/\tau$:

1. $F(100) \approx 10^2/10^7 = 10^{-5}$

2. $F(2{,}500) \approx 2.5 \times 10^3/10^7 = 2.5 \times 10^{-4}$
3. $F(10{,}000) \approx 10^{-3}$
4. $F(50{,}000) \approx 0.005$

(Note that the negative exponential distribution is the only one which does not take account of ageing, so that the reliability of a device is independent of its period of survival and dependent only on its period of 'further' survival.)

4.4 SYSTEMS RELIABILITY

Large-scale systems consume large amounts of money. It is reasonable, therefore, to expect a successful record of operation from them, which is another way of saying a high degree of reliability. Alternatively, large-scale systems usually serve a very large population which comes to depend on it, e.g. electricity supply, telephone service, national defense. The system must be utterly dependable, for without it, even for short periods, users can be placed at great inconvenience and possibly danger.

'You get what you pay for' is a cliché that has a grain of truth in it in relation to reliability. Cheap items can be more prone to failure than the expensive variety (of the same type of course). However, there is 'safety in numbers' and redundancy, the paralleling of items to grant security to a system, is an integral consideration in reliability engineering. This leads us to conclude that the achievement of reliability is not a straightforward operation. In a large-scale system, subsystems for which may be electrical transmission lines or major telephone switching centers, overall systems reliability is made up of a widespread distribution of reliabilities. The subsystems compete for resources to satisfy their individual reliability standards and this is where the systems team must exercise due control to achieve an overall optimum.

Before thinking of how an optimal balance may be achieved, it is first necessary to ascertain how in the arrangement of subsystems to form an overall system, the reliability of that system is affected by individual subsystem reliabilities. We shall now consider, therefore, the reliability laws which govern the composition of systems, which are analogous to the probability laws expressed in the preceding chapter. The use of the conditional probability law will be of great value.

Taking first the simple case of Figure 4.4, by what arguments is it possible to express R, system reliability, in terms of R_A and R_B? We have

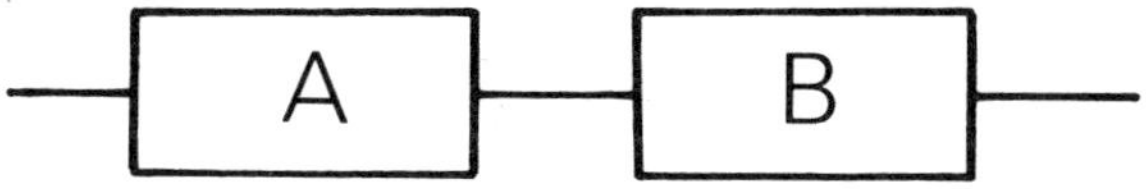

Figure 4.4

said that reliability is a probability. A probability of what? In fact it is said to be the probability that the system will operate as required. Clearly for our series system to work, both A and B must operate as required. Then R is the probability that both A and B operate as required; in other words, it is the event probability of two independent events being therefore the product of the two. Therefore:

$$R = R_A \times R_B \tag{4.24}$$

This assumes that the failure of either A or B will not affect the performance of the other. In other words, successful operation of either A or B is unaffected by the failure of either. This result can be derived more formally by an approach that has extensive application and is based on the probability axioms summarized in terms of the conditional probability law.

If an event is conditioned by other events, e.g. the system's operation depending on whether A operates or not, the strict interpretation of its event probability is given by:

$$p(\mathrm{A}) = \Sigma_{i=1,n}\ p(\mathrm{A}|\mathrm{B}_i)\cdot p(\mathrm{B}_i) \tag{4.25}$$

The B_i are the conditional events and A is the event being considered. The terms $P(\mathrm{A}|\mathrm{B}_i)$ are transitional (or conditional) probabilitics of A happening given B_i has happened. The simple case of equation (4.25) is:

$$p(\mathrm{A}) = p(\mathrm{A}|\mathrm{B}_1)\cdot p(\mathrm{B})$$

and the next to consider is:

$$p(\mathrm{A}) = p(\mathrm{A}|\mathrm{B}_1)\cdot p(\mathrm{B}_1) + \mathrm{p}(\mathrm{A}|\mathrm{B}_2)\cdot p(\mathrm{B}_2) \tag{4.26}$$

The two terms on the right-hand side refer to mutually exclusive events. This will be so if events B_1 and B_2 are mutually exclusive. This simple form of the conditional probability law can now be written so as to reflect system reliability as follows:

$$\begin{aligned} p(\text{system up}) &= \text{reliability of system} \\ &= p(\text{system up}|\text{A is up})\cdot p(\text{A is up}) \\ &\quad + p(\text{system up}|\text{A is down})\cdot p(\text{A is down}) \end{aligned} \tag{4.27}$$

Applying this to the system of Figure 4.4 gives

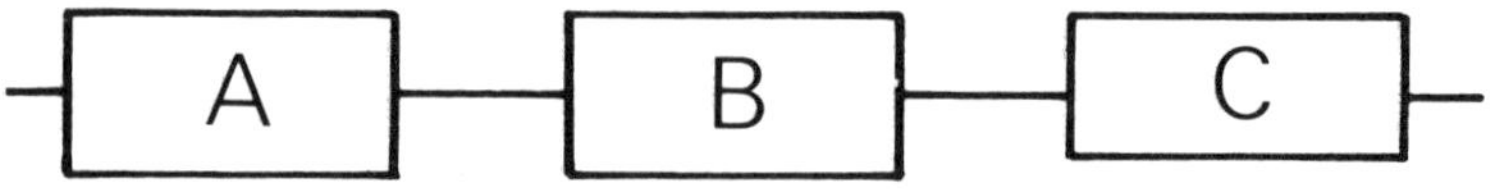

Figure 4.5

$$R = p(\text{system up}|\text{A is up})\, R_A + p(\text{system is up}|\text{A is down})(1 - R_A)$$

The second of the two conditional probabilities is easy to form – it is zero. The other depends on what B is doing and is, in fact, R_B. Therefore we obtain exactly the same result as before. By a process of induction, this result can be extended to N subsystems in series. Moving to three A, B, C (Figure 4.5):

probability that system is up
= reliability, R
= probability (that system is up|C is up)·p(C is up) +
probability (that system is up|C is down)·p(C is down)

The first conditional probability is the reliability of a system with A and B (in series) only which we have found to be $R_A R_B$. The second is zero again. Therefore $R = R_A R_B R_C$. A different geometry is the arrangement of subsystems in parallel. This would normally introduce redundancy, for only one of N in parallel would need to function for the parallel arrangement to perform its function in the system. The reliability increases as redundancy increases. Consider A and B in parallel (Figure 4.6).

probability system is up
= system reliability, R
= probability (system is up|A is up)·p(A is up)
+ probability (system is up|A is down)·p(A is down)

The first conditional probability is one. The second is R_B, therefore:

$$R = 1 \cdot R_A + R_B(1 - R_A) = R_A + R_B - R_A R_B$$

This suggests the probability law for non-mutually exclusive but independent events, namely:

$$p(A \cup B) = p(A) + p(B) - p(A \cap B)$$

or

$$p(\text{A or B works, i.e. system up}),\ R = R_A + R_B - R_A R_B \qquad (4.28)$$

The reader should now attempt to show that for three subsystems in

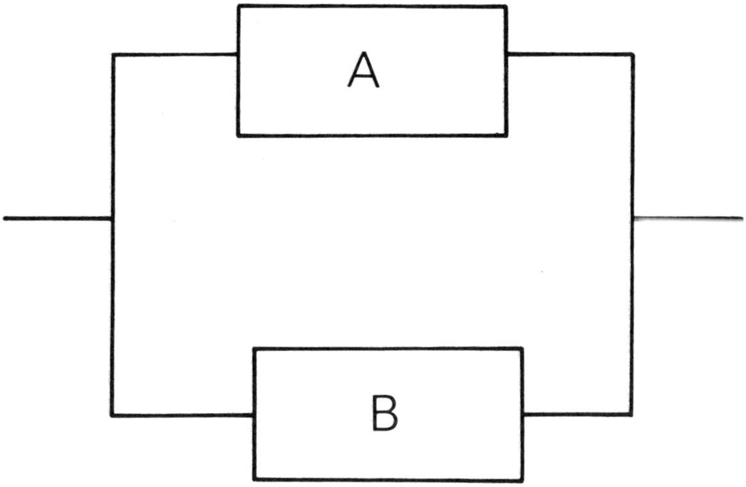

Figure 4.6

parallel with respective reliabilities of R_A, R_B and R_C, the system reliability R is given by equation (4.29) on the basis that the system will operate if one or more of the subsystems operate:

$$R = R_A + R_B + R_C - R_A R_B - R_A R_C - R_B R_C + R_A R_B R_C \quad (4.29)$$

This type of system is described as fully redundant, i.e. only one of its subsystems need be up, though more than one may be, in order for the system itself to be up. In this case a different approach to the derivation of system reliability, R, may be taken.

The unreliability of the system, u, is the product of the subsystem unreliabilities. So for n subsystems, then

$$u = u1 \cdot u2 \cdot u3 \ldots un \quad (4.30)$$

Since $R = 1 - u$ and $R_i = 1 - u_i$, this can be written as:

$$R = 1 - \Pi_{i=1,n} (1 - R_i) \quad (4.31)$$

which is a considerably more succinct way of describing the system reliability.

In some instances total redundancy is not permissible. For example, on the wide-bodied jet aircraft such as the Boeing 747, four identical engines are provided but more than one of these must be operating in order for the aircraft to take off. Surprisingly it needs only one to be operating for it to land. In this case a partial redundancy analysis is called for. When subsystems are identical, in which case the reliability can be said to be constant, independent and the same for each subsystem, the analysis is extremely straightforward and calls for the use of the binomial probability distribution. We now see the system, composed of n subsystems, as a series of outcomes to an experiment in which each outcome may be either successful or unsuccessful. To consider the system as a whole to be successful is equivalent to finding a certain minimum number (or more) of successful outcomes. In general terms, if p is the subsystem reliability, paraphrased as the probability of a successful single outcome and k the minimum number of subsystems needed to be operating successfully in order for the system as a whole so to be operating, then the system reliability R is given by:

$$R = B(n,p,j \geqslant k) \tag{4.32}$$

Explicitly this is

$$R = \Sigma_{j=k,n}\, nCj\; p^j\, (1 - p)^{n-j} \tag{4.33}$$

The reader should note that in the case of total redundancy, then $k = 1$, and we get

$$R = B(n,p,j \geqslant 1) = 1 - B(n,p,0)$$
$$R = 1 - (1 - p)^n \tag{4.34}$$

confirming the previous result. Consider now the following examples in which the binomial probability distribution function is used to calculate system reliabilities.

Example 4.6

A system consists of four components in parallel. System success requires that at least three of these components must function. Determine (a) the reliability of the system if the component reliability is 0.9 and (b) the reliability of the system if five components are placed in parallel to perform the same function.

Defining R to be system reliability, then:

(a) $R = B(4, 0.9, k > 3) = B(4, 0.9, 3) + B(4, 0.9, 4)$
$= 4(0.9)^3(0.1) + (0.9)^4$
$= 0.9477$

(b) $R = B(5, 0.9, k > 3) = B(5, 0.9, 3) + B(5, 0.9, 4) + B(5, 0.9, 5)$
$= 10(0.9)^3(0.1)^2 + 5(0.9)^4(0.1) + (0.9)^5$
$= 0.99144$

Example 4.7

A standby generating system consists of three nominally identical but independent diesel alternator sets. The emergency power requirement can be met if any two of the diesel sets starts successfully on demand. (a) If the probability of failure to start for each set is 0.04, what is the value of system reliability, R? How is R affected if the failure to start per set is (b) 0.1 and (c) 0.6?

(a) $R = B(3, 0.96, k > 2) = B(3, 0.96, 2) + B(3, 0.96, 3)$
$= 3(0.96)^2(0.04) + (0.96)^3$
$= 0.995328$

(b) $R = B(3, 0.9, k > 2) = 3(0.9)^2(0.1) + (0.9)^3$
$= 0.972$

(c) $R = B(3, 0.4, k > 2) = B(3, 0.4, 2) + B(3, 0.4, 3)$
$= 3(0.4)^2(0.6) + (0.4)^3$
$= 0.352$

Example 4.8

An illuminated mimic diagram in a plant control room has 150 nominally identical bulbs which are required to be permanently illuminated. If the probability of any one bulb being out at any one time is 0.01, what is the probability of the following:

(a) At least five bulbs being out?
$\{B(150, 0.01, k \geqslant 5) = 1 - B(150, 0.01, k < 4)\}$
(b) Not more than three bulbs being out?
$\{B(150, 0.01, k \leqslant 3)\}$
(c) Exactly four bulbs being out?
$\{B(150, 0.01, k = 4)\}$

Therefore:

(a) 0.018 (b) 0.935 (c) 0.047

Since n is large and p is small, we may use Poisson to recalculate. The value of $\lambda (= np)$ is 1.5.

(a) $p(1, 5, k \geqslant 5) = 0.019$

(b) $p(1, 5, k \leq 3) = 0.934$
(c) $p(1, 5, k = 4) = 0.047$

Example 4.9

Calculate the system reliability, R, for Figure 4.7 given that stage 3 needs at least two items to be up in order to operate, whereas assembly A needs only one. Stage 2, to operate successfully, needs only one of its subsystems to be up.

First

$$R = R_1R_2R_3;\ \text{also}\ R_1 = 0.99$$

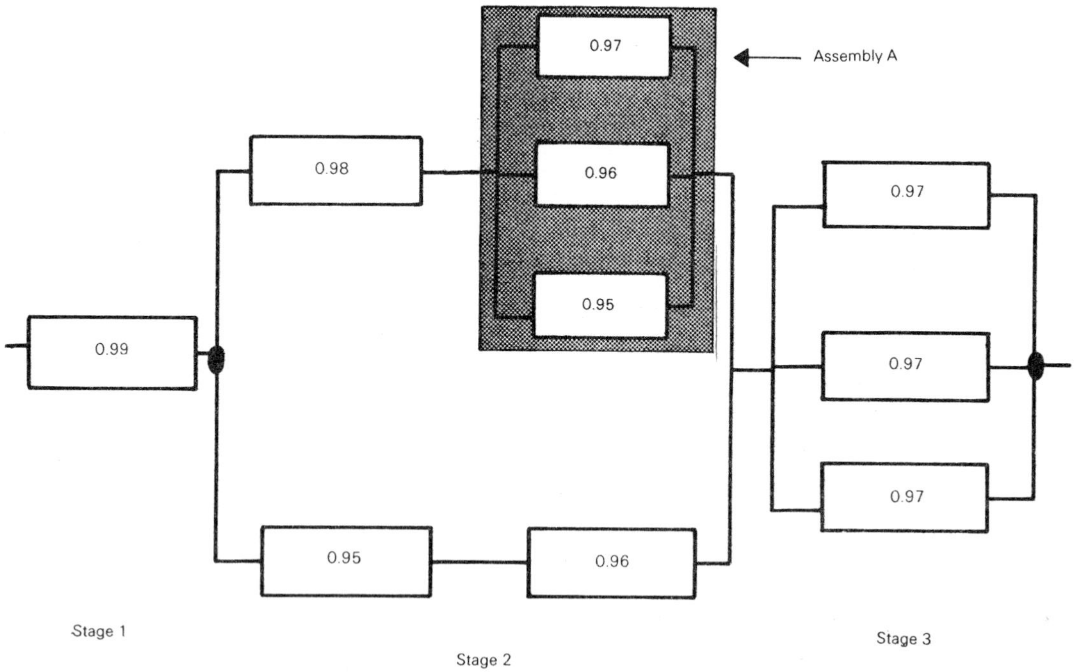

Figure 4.7

$$R_3 = B(3, 0.97, k > 2) = B(3, 0.97, 2) + B(3, 0.97, 3)$$
$$= 3(0.97)^2(0.03) + (0.97)^3$$
$$= 0.997354$$

Now

$$R_A = 1 - (0.03)(0.04)(0.05) = 0.99994$$

Therefore

$$R_2 = 1 - (1 - 0.9799412)(1 - 0.912)$$
$$R = 0.998234826$$

The calculation of the system reliability for partial redundancy when the subsystems do not have identical reliabilities is somewhat more involved but can be found, again from the binomial, this time using the theorem rather than the probability distribution function. This requires an explicit evaluation of the expression

$$\Pi_{i=1,n} (p_i + q_i) \qquad \text{(identically unity)}$$

where individual p_i are subsystem reliabilities. Those terms having at least k contributions of the p_i are aggregated in the final calculation of R.

4.5 RELIABLE SYSTEM DESIGN

From our discussions so far two distinct aspects to reliability evaluation emerge. First, there is the modeling of a system in terms of reliability. This is the problem of producing a suitable geometry (arrangement of subsystems) or family of geometries, that conveys a satisfactory detailed impression of how the system input is transported to become the system output. This modeling relies on a suitable definition or design of the engineering system. This is refined to single out those pieces which demand attention for reliability evaluation. The model or set of models will form the reliability image of the large-scale system itself. The systems engineer is, of course, concerned with several different images of the same system, e.g. traffic, competition and optimization.

A second significant aspect, that differs from the image is concerned more specifically with performance. For a settled, or agreed, geometry the problem often arises of how to allocate a given resource to award various parts of the geometry extra redundant items. This should be attempted in an optimal fashion and to keep things simple, a reasonable objective is to reach the lowest point on the cost v. reliability curve. A variety of design exercises directed at achieving an economic balance between the system reliability and the financial investment in a system will be considered shortly. First, however, the reader should consider the following trial example in which the system reliability should be calculated and some estimate made of the preferred areas to which additional financial resources might go.

4.5.1 Trial example

Example 4.10

An electricity supply system consists of two generator sets, G1 and G2, operated in parallel. The reliabilities of these sets are 0.95 and 0.98 respectively and their costs (per unit) are 183 and 200. Next in series is a transformer T1 having a reliability of 0.99 and a cost of 195. A double circuit line connects T1 to T2, a second transformer which feeds an area of load. The cost of T2 is 176 and its reliability is 0.98. The reliability of a single transmission line is 0.96 and each line can be assumed to fail independently. The cost of the line is 230. A reliability model for this system is shown in Figure 4.8.

Assuming G1 and G2 to fail independently and that system stage 1 is deemed to operate satisfactorily if one or both sets operating, then:

$$\begin{aligned} R_1 &= 1 - (1 - 0.95)(1 - 0.98) \\ &= 1 - (0.05)(0.02) = 1 - 0.001 = 0.999 \end{aligned}$$

$$R_2 = 0.99$$

$$\begin{aligned} R_3 &= 1 - (1 - 0.96)(1 - 0.96) \\ &= 1 - (0.04)^2 = 1 - 0.0016 = 0.9984 \end{aligned}$$

$$R_4 = 0.98$$

Now system reliability R is equal to the product of subsystem reliability,

$$R = (0.999)\cdot(0.99)\cdot(0.9984)\cdot(0.98) = 0.9676789$$

The weakest link in the chain is T2 and as it happens, this is also the cheapest. By investing in a second T2 the system reliability is increased by 2 percent at a cost of 176. By comparison a second T1 increases R by 1 percent at a cost of 195.

It should be very quickly become obvious that if a system designer is to make any practical kind of evaluation of systems reliability against a

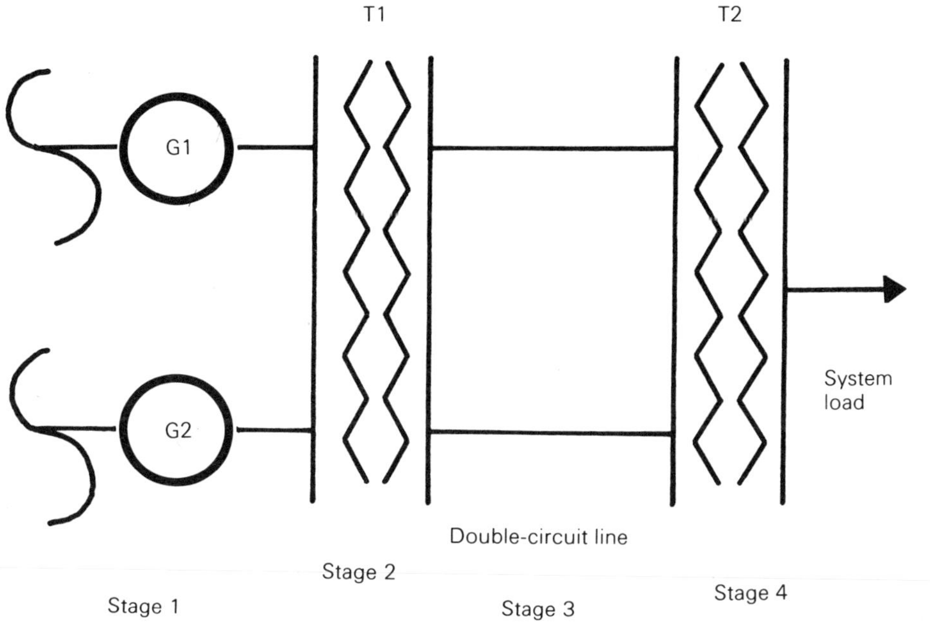

Figure 4.8

financial investment, having regard for all other technical factors, then some kind of formal scheme is required as a design aid. The following exercises are merely example procedures of how such a formal scheme could be arrived at.

4.5.2 Reliability design and optimization

We consider three general situations where an economic balance must be struck between the system reliability and the financial investment. Each is a very elementary example which is intended quite simply to provide a general appreciation of how reliability engineering can affect system design. The following titles suggest the individual objectives in each situation:

1. Optimal design configuration of a series/parallel system minimizing reliability subject to a budgetary constraint.
2. Optimal design configuration of a series/parallel system subject to constraints on safety and budget levels.

3. Optimal mix of redundancy/design effort to minimize the total cost of achieving at least a specified level of reliability.

Minimizing reliability subject to a budgetary constraint

Turning to the first of these three we note that in the general case the system will consist of N stages in series and is, of course, only operable if each stage is operating successfully. To increase system reliability, components can be reproduced, in parallel, in any stage. This procedure is standby redundancy, it being assumed that the stage will operate successfully when only one component is so operating. (In our earlier analysis we referred to this case as total redundancy.) Given the individual component reliabilities and costs, the problem is to determine that configuration which maximizes system reliability subject to a limit on budget for the system. Let the variable denoting the number of components in the kth stage be nk and the unit reliability and unit costs of components used in this stage be pk and ck respectively. Assume the budget limit to be B. Then the system reliability, R, is given by:

$$R = \Pi_{i=1,N}\,(1 - (1 - p_i)\hat{\ }n_i) \qquad (4.35)$$

The cost of the system, C, is given by:

$$C = \Sigma_{i=1,N}\, c_i n_i$$

The design problem, according to this mathematical reliability model, is to choose suitable values for the n_i, $i = 1, \ldots, N$, such that $C \leqslant B$ and R is a maximum value.

Example 4.11

A system comprises three stages. Components can be reproduced in parallel in the first two stages. There can only be one component in the third stage. The probabilities that, at any time, a component is in a functioning stage in stages 1, 2 and 3 are $p_1 = 0.9$, $p_2 = 0.7$ and $p_3 = 0.1$ respectively. The respective probabilities of a component being in a failed stage are: $q_1 = 0.1$, $q_2 = 0.3$, $q_3 = 0.1$. The component costs are $c_1 =$ £2,000, $c_2 =$ £3,000, $c_3 =$ £1,000. The total budget $B =$ £10,000. Using these data and the model above, we get the statements:

$$\begin{aligned}\text{Maximize } R &= (1 - q_1\hat{\ }n_1)(1 - q_2\hat{\ }n_2)(1 - q_3\hat{\ }n_3)\\ &= (1 - 0.1\hat{\ }n_1)(1 - 0.3\hat{\ }n_2)(1 - 0.1\hat{\ }n_3)\\ &= 0.9(1 - 0.1\hat{\ }n_1)(1 - 0.3\hat{\ }n_2)\end{aligned}$$

Subject to

$$n_1c_1 + n_2c_2 + n_3c_3 \leqslant B$$

or

$$n_1 2{,}000 + n_2 3{,}000 + 1{,}000 \leqslant 10{,}000$$

or

$$2000n_1 + 3{,}000n_2 \leqslant 9{,}000$$

The factor of 0.9 in the objective function can be ignored and so the problem in its simplest form is maximize:

$$R = (1 - 0.1\hat{}n_1)(1 - 0.3\hat{}n_2)$$

subject to

$$2n_1 + 3n_2 \leqslant 9;\ n_1, n_2 \geqslant 1.$$

For this simple exercise, the optimal values of n_1 and n_2 can be obtained by trial and error:

1. Let $n_1 = 1$, then $2 + 3n_2 \leqslant 9$ will mean that $n_2 \leqslant 2.33$; therefore let $n_2 = 2$. If $n_1 = 1$, $n_2 = 2$, $R_s = 0.737$ and cost $= 9$.
2. Let $n_1 = 2$, then $4 + 3n_2 \leqslant 9$ will mean that $n_2 \leqslant 1.66$; therefore let $n_2 = 1$. If $n_1 = 2$, $n_2 = 1$, $R_s = 0.624$ and cost $= 8$.
3. Let $n_1 = 3$, then $6 + 3n_2 \leqslant 9$ will mean that $n_2 \leqslant 1$; therefore let $n_2 = 1$. If $n_1 = 3$, $n_2 = 1$, $R_s = 0.629$ and cost $= 10$.

Any other combination of n_1 and n_2 will result in the budgetary constraint being exceeded. Thus the maximum reliability of the system occurs with one component in the first stage and two components in the second stage.

When considering the possible benefits to be derived from redundancy of highly reliable components, it should be noted that often very little is obtained for the extra cost of redundancy. This is demonstrated in the above example when the components are placed in the first stage, rather than two, and yet the reliability of the system is increased only by 0.005 for an extra £1,000 in equipment costs. The solution method adopted for the above example is adequate for simple problems. For more complex problems relating to series/parallel systems, dynamic programming can be used (Chapter 7).

Consideration of budget and safety levels

In some instances, e.g. electronic equipment, if no controlled input signal is made, the possibility of a spurious signal being picked up and triggering off the operation of a system component, and hence setting other stages into operation, increases as redundancy increases. There can be a need, therefore, to examine the problem of determining the optimal design configuration subject not only to a budgetary constraint, but also to a safety constraint in terms of the probability of premature operation of the equipment. We now look at one such situation.

The equipment configuration is as before, where system reliability is given by:

$$R = \Pi_{i=1,N} (1 - (1 - p_i)\hat{\ } n_i)$$

It is assumed that premature operation of the equipment would occur if a component in the first stage operated as a result of picking up a spurious signal, and this component then triggers off the remaining stages. The probability of a component in the first stage operating in the absence of a controlled input signal is, say, p. The probability of the equipment operating when there is no controlled input signal is: probability first stage operates × probability second stage operates × . . . probability Nth stage operates.

Now, probability of a component in the first stage picking up a spurious signal is p. Therefore the probability of a component not picking up the signal is $q = 1 - p$. Therefore: probability of first stage not functioning due to a spurious signal is $q\hat{\ } n_1$. Probability of first stage operating due to a spurious signal is $1 - q\hat{\ } n_1$. Probability of system operating due to a spurious signal when there is no controlled input signal is

$$(1 - q\hat{\ } n_1)\ \Pi_{i=2,N} (1 - (1 - p_i)\hat{\ } n_i)$$

The safety constraint on the design configuration is that the probability of premature operation of the system due to a spurious signal must be less than or equal to some predetermined factors, s. Thus:

$$(1 - q\hat{\ } n_1)\ \Pi_{i=2,N} (1 - (1 - p_i)\hat{\ } n_i) \leqslant s$$

The budgetary constraint is the same as before:

$$\Sigma_{i=1,N}\, c_i n_i \leqslant B$$

The problem is now to determine the optimal number of components in parallel in each stage of the system in order to maximize R, given by

$$R = R = \Pi_{i=1,N} (1 - q_i\hat{\ } n_i)$$

subject to

$$(1 - q\hat{\ }n_1)\ \Pi_{i=2,N}\ (1 - (1 - p_i)\hat{\ }n_i) \leqslant s$$

and

$$\Sigma_{i=1,N}\ c_i n_i \leqslant B$$

This is a model of the problem relating component numbers n_i to equipment reliability, safety and cost.

Example 4.12

The data are as for the previous example (Table 4.2), $N = 3$, $B = £10{,}000$ in addition to which we let $q = 0.95$ and $s = 0.075$.

Table 4.2

i	q_i	$c_i(£)$
1	0.1	2,000
2	0.3	3,000
3	0.1	1,000

In order not to exceed the budgetary constraint we know from the previous example that the possible values of $n1$, $n2$ and $n3$ are restricted to three situations which can be summarized in Table 4.3 which includes also the safety level for each case. The last of these three does not meet the safety level and of the remaining two, the configuration $n1 = 1$, $n2 = 2$, $n3 = 1$ is that which gives the optimal design.

Table 4.3

System configuration $n1$	$n2$	$n3$	System reliability	Safety level
1	2	1	0.737	0.041
2	1	1	0.624	0.061
3	1	1	0.629	0.090

The interested reader may wish to consult Dummer and Winton (1974) which discusses problems of premature operation of series/parallel systems. In this example it has been assumed that premature operation of the system could occur if one component in the first stage operated because of a spurious input signal. To reduce further the probability of premature operation, it might be possible to design the stage so that at least r of the $R1$ components would have to operate before a signal would be passed on to the next stage. This is termed voting redundancy.

Optimal mix of redundancy/design effort

In our final design situation we look at the trade-off between design effort and redundancy in achieving at least a specified level of reliability.

Equipment composed of one stage is required to be designed to operate for one unit of time at a reliability at least equal to that specified by the user. The design team can meet this requirement through the use of standby redundancy within the stage or by designing equipment which can meet the reliability requirement without using redundancy. The design of such reliable equipment is costly and a balance is required between the amount of money spent in design and the use of redundancy.

Let p_j be the reliability of a component of type j, where j takes values 1, 2, . . ., m. Component type 1 is less reliable than type 2, type 2 being less reliable than type 3, etc., i.e. $p_{j+1} > p_j$ for $j = 1, 2, \ldots, m - 1$.

The reliability, R, of equipment composed of one stage, having n_j components of type j in parallel is given by:

$$R = 1 - (1 - p_j)\hat{\ }n_j$$

The design and operation cost of a component of type j for one unit of time (for which p_j is defined) is $A + B_j$, where A and B are constants. The fixed cost of a component is F, independent of design specification. The equipment reliability must equal at least R_t, some predefined level. The problem is to determine the optimal type of component j to design, and the optimal number of these to have in parallel in the system in order to minimize the total system cost over one unit of time, subject to a reliability constraint. Thus we wish to minimize C, where

$$C = n_jF + (A + B_j)\, n_j$$

$j = 1, 2, \ldots, m$ subject to $R > R_t$.

This is a model of the problem relating type of component j, and number of components of type j in parallel, n_j, to total cost and system reliability.

Example 4.13

The reliabilities of component types (and unreliabilities) are summarized in Table 4.4. Take $F = 1{,}000$, $A = 500$ and $B = 1{,}000$. Assume R_t to be 0.95. Thus, from the model, we wish to minimize C, where

$$C = n_j * 1{,}000 + (500 + 1{,}000j) * n_j$$

$j = 1, 2, \ldots, 5$ subject to $R \geqslant 0.95$.

Table 4.4

j	p_j	q_j
1	0.75	0.25
2	0.8	0.2
3	0.85	0.15
4	0.9	0.1
5	0.95	0.05

Table 4.5

j	n_j	R_e	Cost (£)
1	3	0.98	7,500
2	2	0.96	7,000
3	2	0.98	9,000
4	2	0.99	11,000
5	1	0.95	6,500

By simple calculation we summarize the results of this example in Table 4.5. From this we see that the least expensive situation is to purchase the most expensive item (and most reliable), and have only one of this type. However, if our value of R had been say, 0.975, the optimal solution would then be to have three of the cheapest type.

An obvious extension of the problem is to systems composed of several stages. Amstandter (1971) discusses the use of equipment over several periods of time and includes increasing operating costs per unit of time in the model. Selection of the optimal design configuration is then based on the total cost of design and operation over the period for which the system is required. An exponential trend in design costs is used rather than a linear trend to give more reality to the model.

4.6 RENEWAL AND/OR REPLACEMENT POLICIES

We consider now, very briefly, a complementary aspect to system reliability, but one which nevertheless deals with failure in systems. This is the subject of renewal and it is one that has received a great deal of attention in the relevant literature. Some of the more important references are given at the end of this chapter. We shall introduce the subject in outline only by a brief discussion of one elementary exercise.

As a component enters its final phase of operating life, the wear-out phase, it is reasonable to think in terms of replacement. The problem is to decide the best policy to adopt with regard to replacement. The replacement should be made when the new equipment offers more economical or efficient service than the old. This might happen for a number of reasons, e.g. the old equipment has failed and does not work at all; the old equipment has deteriorated and works badly or requires expensive maintenance; a better design of equipment has been developed; the old equipment is expected to fail shortly.

A great deal of study has gone into the formulation of policies based on a variety of models and theories. As an illustration we consider a best policy for replacing items of a large collection which do not so much deteriorate gradually but rather fail without prior warning expensively.

Suppose there is a large number of items, all of which must be kept in working order. If one fails in service it costs 1 unit to replace, but if all are replaced in the same operation, it costs 0.35 units per item. If the proportion of items failing in successive time intervals is known, it is then possible to decide on a suitable replacement policy. The given distribution of item lives is shown in Table 4.6. We assume that all items failing during a given week do so just before the end of the week and that group replacements can only be made at the end of the week. Further, that the actual percentage failures are the expected percentage failures based on the probability distribution. What we require to know, therefore, is the

Table 4.6

Week no.	Proportion failing
1	0.09
2	0.16
3	0.24
4	0.36
5	0.12
6	0.03

Table 4.7

Replace after week	Proportion failed in week	Total proportion of failures to date
1	0.09	0.09
2	$0.16 + (0.09)^2 = 0.17$	0.26
3	$0.24 + 2(0.09 \times 0.16) + (0.09)^3 = 0.27$	0.53
4	$0.36 + 2(0.09 \times 0.24) + (0.16)^2 + 3(0.09)^2\,0.16 + (0.09)^4 = 0.43$	0.96

cost/item/week of replacing at the end of one week, two weeks, etc. All failures will be replaced but the decision remains as to when to replace an entire population.

The numerical discussion is as follows. The proportion failed after one week is 0.09. The population at the start of the second week consists of 0.91 of a week old and 0.09 of zero weeks old. The replacements of last week will suffer a proportional loss after one week of 0.09 so the total replacements after two weeks is $0.16 + (0.09)^2$. Both these constitute subpopulations that will start to fail in proportions shown in Table 4.7. After three weeks the original population will suffer a proportion failure of 0.24. The subpopulation starting after the first week will suffer a 'second week' population failure of 0.16. The two subpopulations starting at the end of the second week will both suffer a 'first week' population failure of 0.09. Therefore at the end of the third week the population failure will be

$$0.24 \times 2(0.09)(0.16) + (0.09)^2$$

The number of subpopulations is now four. Two of these will suffer 'first week' population proportion failures, one will suffer 'second week' population proportion failures and one will suffer a 'third week'. Eight new subpopulations will be introduced at the end of the fourth week.

Replacing after week 1 will cost 0.35 units per bulb. Replacing after week 2 there will be the added cost of having replaced total population of failures to date of previous week. The average cost of replacing after various weeks is given in Table 4.8, where it is possible to see that group replacement after the third week is the optimal policy since it is the minimum average cost of group replacement. It is also cheaper than not having any group replacement policy. This latter cost is worked out by dividing the cost of replacing a bulb by the average life of a bulb, which is

Table 4.8

Replace after week	Total cost per bulb	Average cost (per week)
1	0.44	0.350
2	0.61	0.305
3	0.88	0.293
4	1.31	0.327

$$1 \times 0.09 + 2 \times 0.16 + 3 \times 0.24 + 4 \times 0.36 + 5 \times 0.12 + 6 \times 0.03 = 3.35 \text{ weeks}$$

The cost therefore is $1/3.35 = 0.30$ units.

A general analysis of this problem shows the binomial expansion of subpopulation and would give an indication of how complex decision-making can be made. Provided a population has not expired, i.e. it is within six weeks of being derived from an original population, it will be transformed into two populations by virtue of having a proportion fail in the next week and the remainder survive to be one week older. Therefore:

number of populations at stage k = number of populations at stage $k - 1$ + new population

new populations at stage k = number of populations at stage $k - 1$

whence

number of populations at stage k
= 2 × number of populations at stage $k - 1$

There is, therefore, a doubling of number of populations with discrete jumps in cycles of six weeks. An initial six weeks' cycle can be described by Table 4.9, which shows the distribution of population ages. A population introduced at the start of the week is 0.

4.7 MARKOV PROCESSES

We conclude this chapter on systems reliability with just a few brief remarks on an important class of processes known as Markov chains or Markov processes. Mathematical models on Markov processes have found wide applicability in the fields of theoretical physics, astronomy, biology and the social sciences. Additionally, many reliability problems

Table 4.9

Start of week	Population distribution
1	0
2	1 0
3	2 1 0 0
4	3 2 1 1 0 0 0 0
5	4 3 2 2 1 1 1 1 0 0 0 0 0 0 0 0
6	5 4 3 3 2 2 2 2 1 1 1 1 1 1 1 1
	0 0 0 0 0 0 0 0 0 0 0 0 0 0 0 0

can be represented in terms of Markov processes and it is appropriate therefore that an outline of their nature be made at this juncture.

First, a stochastic process is a sequence of trials of an experiment, defined in the probabilistic sense. A Markov process is distinguished as a stochastic process in the fact that the outcome on any particular trial is dependent on the outcome of the preceding trial. A conditional probability is introduced in association with every pair of outcomes. We must also introduce state and time concepts. For example, we must define a space of possible states and define how transitions are made over a sequence of trials.

We may, for example define the states of a machine as operating or failed, and consider how transitions are made back and forth from each of the possible states. The following are the classes of systems which can be described by Markov processes:

1. Time and space discrete.
2. Time continuous and space discrete.
3. Time and space continuous.

Many reliability problems are of class 3, but we will confine our attention entirely to class 1. Our analysis of this proceeds as follows.

The system has a given set of states, equivalent to the outcome of an experiment governed by chance, which is characterized by the matrix $(1 \times N)$:

$$x = (x1, x2, x3, \ldots, xN)$$

The system can only occupy one of these states at any point in time, but with the passage of time moves from one state to another or remains where it is. The probability that a system moves from one state to another in the next interval of time depends only on the two states, i.e.

going from i at time t to j at time $t + \Delta t$ is pij. So the system's 'memory' extends back to the previous state only. One can, from this, construct a matrix whose elements are the transitional probabilities. Therefore:

$$
\begin{aligned}
& p11, p12, \ldots, p1N \\
P = \; & p21, p22, \ldots, p2N \\
& \ldots \\
& pN1, pN2, \ldots, pNN
\end{aligned}
$$

Since the system must be in one of the states at any point in time, the rows of this matrix each sum to unity. Such a matrix is known as stochastic, and P is known as the stochastic transitional probability matrix for a discrete stationary Markov process (Markov chain).

The a priori probability that the system occupies state i at any time is pi and, clearly, to describe the system state, the matrix $(1 \times N)$

$$\pi = (p1, p2, \ldots, pN)$$

would be used, in which case the relation between $\pi(t)$ and $\pi(t + \Delta t)$ will be

$$
\begin{aligned}
& \pi(t + \Delta t) = \pi(t) \cdot P \\
& (1 \times N) \; (1 \times N) \; (N \times N)
\end{aligned}
\tag{4.36}
$$

From this the set

$$
\begin{aligned}
\pi(1) &= \pi(0) \cdot P \\
\pi(2) &= \pi(1) \cdot P \\
\pi(3) &= \pi(2) \cdot P \\
&\vdots \\
\pi(N) &= \pi(N - 1) \cdot P
\end{aligned}
$$

is developed for a discrete form. Notice then that

$$\pi(2) = \pi(1) \cdot P = \pi(0) \cdot P \cdot P = \pi(0) \cdot P^2$$

where $P^2 = P \times P$ by matrix multiplication. Whence $\pi(k) = \pi(0) \cdot P^k$

Is Pk a stochastic matrix? The definition of a stochastic matrix is $P^k e = e$ where e is an $N \times 1$ matrix with each of its elements equal to unity. Thus the rows of a stochastic matrix sum to unity. Since P is a stochastic matrix, then $Pe = e$. What about P^2? Well $P^2 = P \cdot P$. Therefore $P^2 e = P \cdot Pe = Pe = e$. Therefore P^2 is stochastic and so is P^k, $k > 2$.

It is conceivable that the process reaches a steady-state whereby transitions are unaffected by previous system performance so that in the long run a system will occupy a state with constant probability: in this case

$$\pi = \pi \cdot P \tag{4.38}$$

so we may solve for π knowing that it is itself a stochastic matrix $(1 \times N)$.

For example, on any given day a power station may be on full load or not, states 1 and 2 respectively, and the load varies such that it can be represented as Markov process with matrix P given as in Table 4.10. This can be represented in diagrammatic form as shown in Figure 4.9.

Table 4.10

		1	2
$P =$	1	0.6	0.4
	2	0.2	0.8

Assuming that the variation is a daily one and on Wednesday the station was on full load, what is the probability that it will be on full load on Saturday? In the long run what is the likelihood it will not be on full load?

Let Wednesday be time zero when $r(0) = (1, 0)$. Saturday will be given by $r(3)$, where $r(3) = r(0)p3$; see Table 4.11.

Therefore the probability that the power station will be on full load on Saturday is 0.376. Let the long run probabilities be $p1$ and $p2$ respectively, then:

$$(p1, p2) = (p1, p2)\begin{pmatrix} 0.6 & 0.4 \\ 0.2 & 0.8 \end{pmatrix}$$

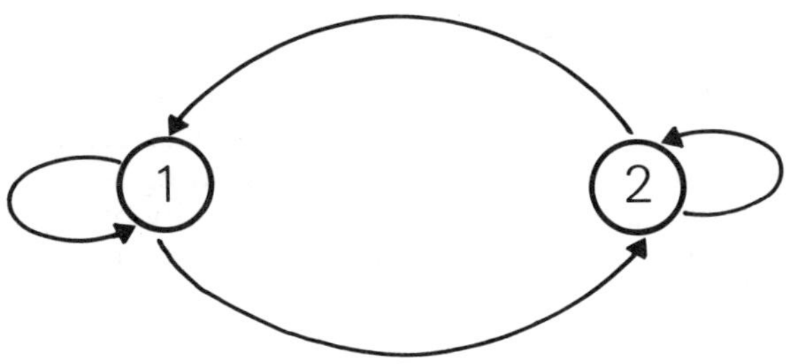

Figure 4.9

Table 4.11

		0.6	0.4		0.6	0.4		0.44	0.56
P^2	=			×			=		
		0.2	0.8		0.2	0.8		0.28	0.72
		0.44	0.56		0.6	0.4		0.376	0.624
P^3	=			×			=		
		0.28	0.72		0.2	0.8		0.312	0.688

or written out:

$$p1 = 0.6\,p1 + 0.2\,p2$$
$$p2 = 0.4\,p1 + 0.8\,p2$$

These are both the same equation

$$p1 = 0.5\,p2$$

and to solve we need to know $p1 + p2 = 1$. Therefore $p1 = 1/3$ and $p2 = 2/3$.

The power station is twice as likely not to be on full load as it is to be on full load. Notice from $p2$ and $p3$ that these terms are the subject of convergence.

It is beyond the scope of this book to consider the further classes of systems to which Markov processes may be usefully applied for reliability evaluation. The reader is referred to the further reading list to continue his study of this subject in the knowledge that he has a good grounding thus far.

4.8 EXERCISES

1. An electrical generating station has a twofold requirement to meet a normal demand of 350 MW and a peak demand of 750 MW. The probability of a normal demand arising at any time is 0.75 and the probability of a peak demand arising at any time is 0.25. The station has two independent generating sets running in parallel and each set has an output of 400 MW. The probability of a generating set being 'off line' at any time is 0.1. What is the reliability of the generating station in meeting its requirement?

2. A continuous variate, x, representing the time to failure of a system, has a probability density function of the following form:

$$f(x) = (\exp(-x/\lambda))/\lambda$$

where λ is a constant of time value 1,000 h.

What is the probability that a particular system will have a time to failure of (a) less than 100 h, and (b) less than 1,000 h?

3. A bank of nominally identical floodlights are used to illuminate a working area. The total number of lights in the bank is n and the probability that any one light is in the successful state at any time is p. What is the mean or expected number of lights that will be in operation at any one time if (a) $n = 20, p = 0.9$; (b) $n = 100, p = 0.9$; (c) $n = 100, p = 0.5$?

4. It is found that a number of system breakdowns occurring in a given length of time follows a Poisson distribution with a mean value of two breakdowns. What are the probabilities, in the same length of time, of the system having exactly (a) no breakdowns, (b) one breakdown, (c) two breakdowns and (d) ten breakdowns?

5. The output of a digital device has a mean value of five units. Random variations cause integer changes in this output which are found to follow a Poisson distribution. If the required output is required to be greater than two units and less than eight units, what is the reliability of the device?

6. One of the processes given below may be described by a binomial distribution, while the other exhibits a Poisson distribution. Give reasons for deciding which process conforms to which distribution law.

(a) The probability of a garbled message along a data link is 5 percent. Find the probability of not more than two garbled messages in 200.
(b) A data link has four channels. The probability of a channel being busy when required is 0.1. What is the probability for immediate access to three channels?

Solve both these problems by the distribution law which you think applies to that problem. Binomial distribution: probability $p(r)$ of r successes in N trials is $(r + 1)$th term of the expansion $(p + q)^n$, where $q = p - 1$. Poisson distribution: in a population n, if the probability of an event is p, then the probability of r events is $p(r)$, where

$$p(r) = \exp(-np) \times ((np)^r))r!$$

7. An inventory of transistors consists of 75 percent from manufacturer X, 99 percent of which give a satisfactory performance. The remainder, from manufacturer Y, are defective in 10 percent of cases. What is the probability that a transistor, selected at random, is made by X and is defective? What is the probability that it is defective irrespective of manufacturer? A transistor is found to be defective. What are the probabilities that it came from X, and from Y? A simple Markov process, with only two outcomes, favorable (F) and unfavorable (U) is shown in Figure 4.10. If the initial probability of a favorable output is $p(0)$, show that the probability of a favorable outcome after n trials is given by:

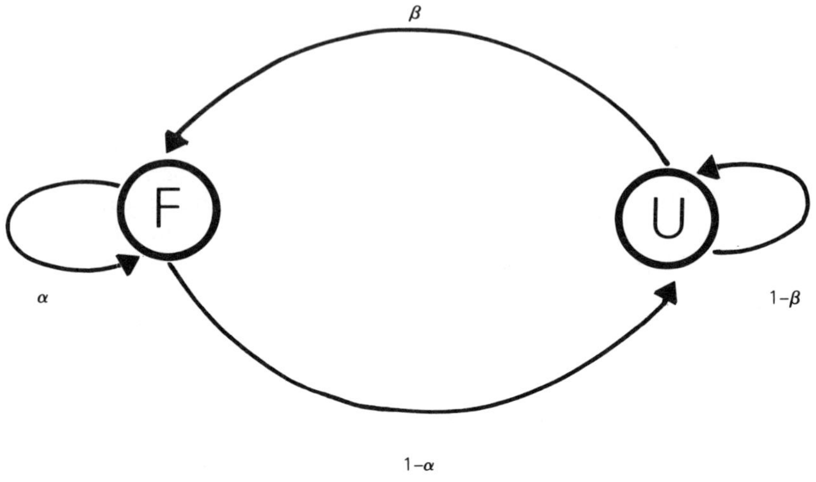

Figure 4.10

$$p(n) = (\alpha - \beta)^n\, p(0) + b(1 - (\alpha - \beta)^n)/(1 - (\alpha - \beta))$$

If α and β are both less than 1, find $p(\infty)$.

8. The probabilities that three missiles will each hit a certain target are 0.5, 0.9 and 0.6 respectively. What is the probability that at least one of them will hit the target?

Determine the reliability of the system shown in Figure 4.11, where the

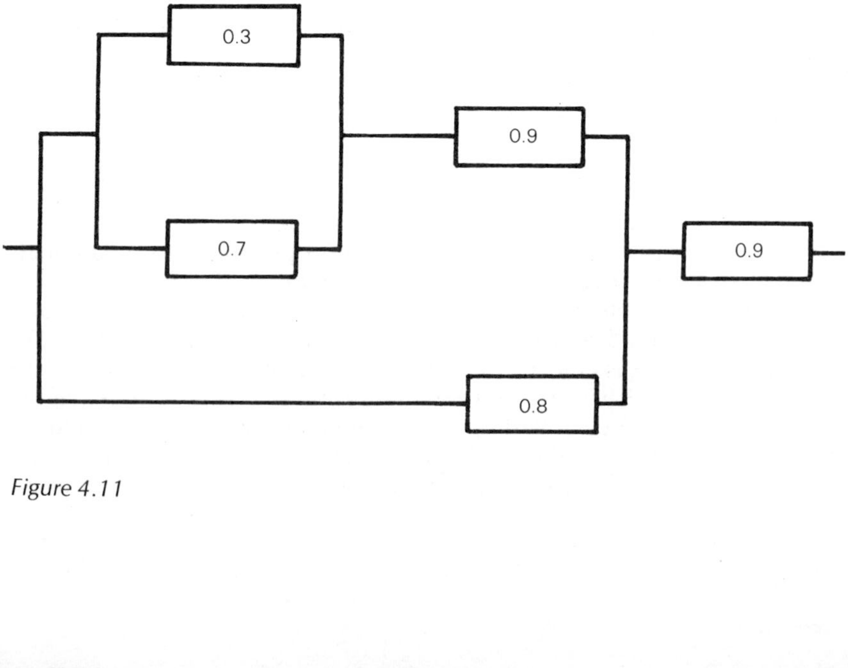

Figure 4.11

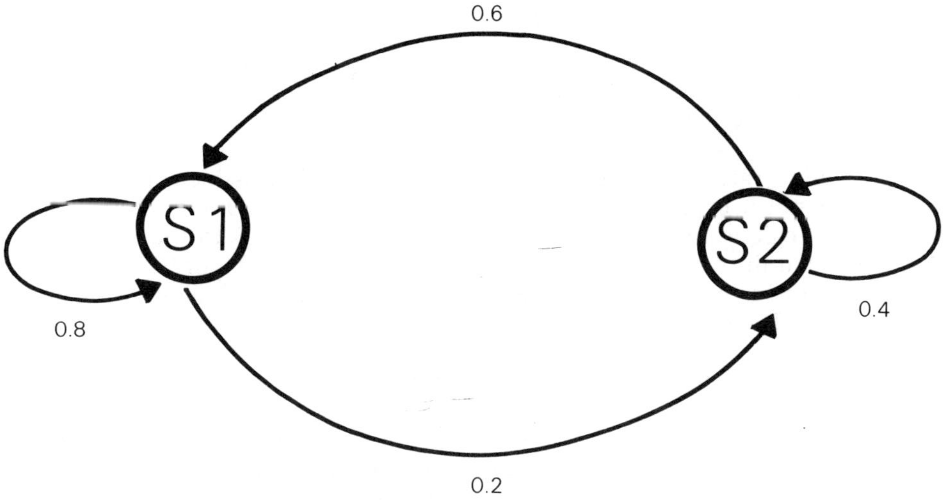

Figure 4.12

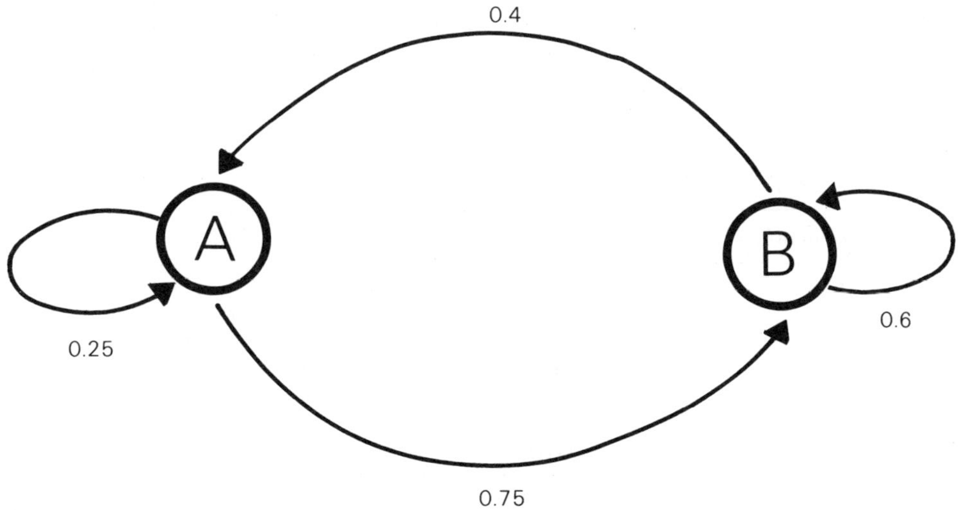

Figure 4.13

probabilities of successful operation of the subsystem are shown. State the probability laws used in determining the reliability.

A missile warning system consists of n components connected in parallel having constant failure rates a_1, a_2, to a_n respectively. Show that the probability that the system will be operating after time T is:

$$R = \Pi_{i=1,n} (1 - \exp(-a_i T)$$

The reliability of a particular system varies daily according to the Markov process shown in Figure 4.12. Show that the system operates correctly (state S1) in the steady-state with probability 0.75.

9. The calling rate, registered at a local exchange, is classified according to whether or not it exceeds the average rate of 150 calls per hour. Class A corresponds to calling rates less than the average and Class B to the remainder. Assuming that the calling rate is a Markov process given by Figure 4.13, find the probability of having a Class A condition in the third hour following an hour when a Class B condition prevails.

10. A particular type of aircraft is fitted with a triplex autopilot system. This comprises three sensibly independent autopilot subsystems each of which is capable of achieving its purpose alone, but all of which normally work together. The complete autopilot provides automatic control for the aircraft elevator, aileron and rudder systems. A block schematic of the triplex elevator control system is shown in Figure 4.14. It has been found that the failure characteristics of each block in this system can be represented by a mean failure-rate function Θ, and that such failures are reasonably independent of each other. Also the failure of any one block results in a failure of the subsystem of which it is a part. The elevator control system is checked for complete working immediately prior to a two hour flight. What is the probability that no automatic elevator control will be available by the end of this flight, if the assumed typical values for Θ are as shown in Table 4.12?

Table 4.12

Item no.	Item description	θ (faults/h)
1	Flight path receiver	10×10^{-4}
2	Radio altimeter	6×10^{-4}
3	Rate gyro	2×10^{-4}
4	Flight controller	3.5×10^{-4}
5	Pitch computer	5.5×10^{-4}
6	Servo amplifier	2×10^{-4}
7	Servo motor	1×10^{-4}

11. A small nuclear research reactor has three absorber rods which are suspended above the reactor and are designed to drop into the reactor core to shut the reactor down in the event of any untoward incident. The three rods are designated A, B and C and it has been found that the probability of each of these rods failing to drop on demand is as follows: $p_A = 0.005$, $p_B = 0.01$, $p_C = 0.001$. If it is known that any two or more of three rods entering the reactor core will safely shut the reactor down, what is the probability of failing to shut the reactor down when called upon?

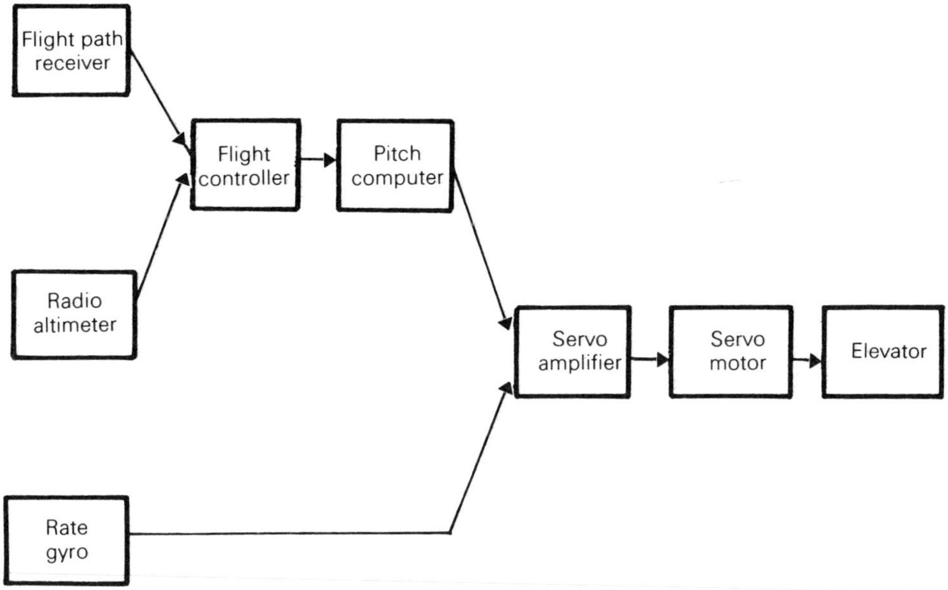

Figure 4.14

12. It is simply assumed that an accident may occur at automatic level crossings as a result of two possible sequence of events:

(a) The complete failure of the automatic mechanism for sounding the alarm and closing the barriers, in which case a train may hit a vehicle which is in the process of negotiating the crossing.
(b) The stalling of a vehicle on the crossing just prior to the automatic closure of the barriers and the inability to move the vehicle before the passing of the train.

On a particular railway system, 100 automatic crossings are in existence and both the arrival of trains and the crossing of road vehicles take place with an equal likelihood at any time. The arrival of trains takes place with a mean rate of one train every two hours and the crossing of vehicles at a mean rate of 104 vehicles/h.

The transit time of road vehicles during the time which they may be at risk has a mean value of 3 s but if a vehicle stalls, then its mean risk time rises to 15 s. One in every million vehicles using the crossing stalls on the crossing, and the probability of complete failure of the automatic system is also one in a million per demand for operation. If all events are assumed to be independent, and the same conditions and parameters pertain over a ten year period, how many accidents would be expected over this period of time?

4.9 FURTHER READING

Amstandter B.L. (1971) *Reliability Mathematics*, McGraw-Hill.

Dummer, G.W.A. and Winton, R.C. (1974) *An Elementary Guide to Reliability*, Pergamon.

Endrenyi, J. (1978) *Reliability Modelling in Electric Power Systems*, Wiley.

Green, A.E. and Bourne, A.J. (1972) *Reliability Technology*, Wiley-Interscience.

Martin, J. (1976) *Engineering Reliability Techniques*, Open University Press.

Pages, A. and Gandran, M. (1986) *System Reliability – Evaluation and Prediction in Engineering*, North Oxford Academic Publishers Ltd.

Tillman, F.A., Ching-Lai Hwang and Way Kuo (1980) *Optimisation of Systems Reliability*, Dekker.

5 Queueing theory

5.1 INTRODUCTION

In Chapter 2 we noted some definitions for systems concepts which are useful in the development of system models, be these mathematical in character or otherwise. Such models facilitate specialized treatments in the system's design based on the distinctive requirements placed upon the system for coping with its inputs.

In the previous chapter we were concerned primarily with the requirements upon the system to ensure a maximum likelihood of the correct things happening to each type of input in order to achieve the desired output. Whilst not examining in any detail the fabrication of a system to achieve the technical processing of inputs, systems reliability is really primarily concerned with the fact that failures in this processing may occur without foreknowledge. Hence it is concerned with a form of establishing decision-making which will account for such failures.

We turn, in this chapter, to an examination of a distinctly different type of requirement upon the system for coping with its input, namely, the organization of technical processing at the proper time of a multiplicity of inputs randomly distributed in time.

Evidence for the existence of this type of input to all kinds of systems, including of course the large-scale variety, is available from everyday life. For example, making purchases of goods, from the point of view of the vendor, involves large numbers of people who make their demands quite randomly. Providing a service to deal either with the selection or delivery of the goods, or taking payment for the goods, requires the vendor to make decisions as to what level of service to provide. By this is meant not merely politeness or efficiency, in themselves important parameters, but also the number of servers to employ and the degree of sophistication in the machinery which servers may have to use, e.g. cash-registers, computer terminals on-line to an information service, petrol pump dispensers, and so on.

Reasons why the vendor should go to such lengths in considering the make-up of the service system should be self-evident. First, the effort he puts into providing an effective service will cost money, which has to be paid for by the users of the service. On the other hand, such users will, in general, be attracted to an effective service, while most people will be discouraged from a service that makes them have to wait, in their view, far too long.

In another example we find that costs are not the only basis for design criteria. The control of air traffic at airports needs a complex organization of people and machines and makes a very suitable subject for systems engineering studies.

The arrival and departure of aircraft at airports, though scheduled, in practice constitute random events. The repercussive effects of delays at other locations, breakdowns of aircraft or terminal services, bad weather and strikes by controllers or ground staff, all contribute to the uncertainty effects of arrivals and departures. From the point of view of the airlines, aircraft on the ground do not make money but rather incur costs, e.g. airport taxes. It is very much in their interests to have the aircraft flying for a high proportion of the time. On the other hand, aircraft in the air must at some stage be brought safely to the ground. Unfortunately the arrival of aircraft into the airspace at airports does not, of necessity, warrant the immediate use of the runway. This may be in use at the time by aircraft taking off or landing, and there are strict rules governing the technical processing of this particular type of input, and there is a limit to the capacity that this service provides. Arrivals of this type must form a queue. The aircraft are placed in stacks by the controllers who inform the pilot to fly in a 'holding pattern'. Clearly there is a limit to how long aircraft may safely wait, but at the same time there is a limit to the controller's ability to cope with congestions of air traffic. Whilst costs are not, by any means, to be neglected in an examination of this type of system, the safety of persons involved is an overriding factor, a point with which air travellers would wholeheartedly agree, even at the expense of personal delays.

A third example is seen in the telephone system. The subscriber in the United Kingdom using STD (subscriber trunk dialing) is interested in obtaining, almost immediately, a dial tone, having taken his receiver off the hook. Further he expects to make his call without having to hear a recorded message telling him 'all lines are engaged – please call later'. For most of the time this expectation will be the case. However, there will be busy periods when his call cannot be placed because the provision of equipment is insufficient to meet the demand during such periods. British Telecom, however, must make decisions as to how much effect to put into the provision of an effective telephone service without making it overly expensive, having consideration for the fact that once equipment is installed, so it must remain even for those long periods when demand is considerably less than that with which it can effectively cope. Although the number of telephone operators which British Telecom needs to employ has been dramatically reduced with the introduction of STD, a small staff is nevertheless still required for some aspects of the telephone service. Here again a decision is required to balance the costs for providing such a staff against the standard of service which users receive in terms particularly of having to wait for an operator to respond to a call.

These then are the kind of problems we shall study as examples in this chapter. Within the general field of operational research, the development of queueing theory as a piece of mathematical research, based on the theory of probability, has been rapid. The systems engineer is fortunate indeed to be able to draw on the large volume of material in this field in order to make the essential decisions in those aspects of his systems design study which basically are related to traffic.

What then are the general features of a traffic problem? Clearly the input is one feature and we need to be able to say something about the nature of this input as an arrival pattern. The service is a second feature. This exists to meet a specific demand, e.g. for air transportation, telephone communication or the purchasing of goods. When localized, the service constitutes a focal point for the technical processing of demand which itself becomes formalized as inputs to the service.

There are two factors which need to be elucidated in describing the nature of the service and these are, first, the rate at which a service operates and, secondly, the influence which it exercises on the inputs whilst these await processing by the service. This latter feature is more properly referred to as queue discipline. We have already stated that arrivals are random events and therefore the number of arrivals are characterized by statistical distributions. Basically these are frequency distributions depicting the number of times given volumes of arrivals occur in a fixed period of time. In many problems, mathematical expressions can be found which give a good enough approximation to the actual distribution, but sometimes the actual distribution, found by observation, must be used.

In a similar fashion, the service time may be specified by a probability distribution since this too must be considered a random variable. Service times will vary according to many factors, included among which are the length of queue and type of item being serviced. Ultimately the value of service time is governed by chance.

Queue discipline describes the behavior of the items as they wait to be serviced. When a queue (or waiting-line) forms, the items may be served in the order in which they arrive, as, for example, shoppers queueing to pay for the goods. Alternatively they may be serviced in an arbitrary manner, as, for example, with telephone calls made upon a telephone operator. A third type of queue discipline can be described as last-come-first-served, as exemplified perhaps by washing dishes which are stacked prior to being cleaned. Finally there may be some kind of priority scheme in operation as, for example, in a dockyard where certain types of cargo, e.g. perishable goods, may need to be unloaded before others. An item is said to have pre-emptive priority if it not merely goes to the head of the queue on arrival but displaces any item already being served. Provided that the order of service is not related to service time, it does not affect the queue length or the average waiting time but it does affect the time an individual unit has a wait.

One final comment about the service should be made. It is generally presumed that the service operates on no more than a single item at any given time. In such cases it should be noted that the service consists of a single service channel. The analysis of queues does allow for multiple service channels operating in parallel and constituting a single service. This is a great help since the parameter, c, denoting the number of channels, can be built into a mathematical model of the traffic system so that, if required, an optimum number of service channels can be determined relative to known traffic conditions and some stated objective function defining optimality.

Before proceeding with the analysis, aimed at a quantitative assessment of waiting-line problems, let us summarize the basic components of any queueing system (Figure 5.1):

1. The input, specified by the type of item and by a probability distribution (or an observed distribution) describing the pattern of arrivals into the system. The pattern may be concerned with the number of arrivals (a discrete random variable) or the time distribution of arrivals, i.e. inter-arrival time (a continuous random variable).

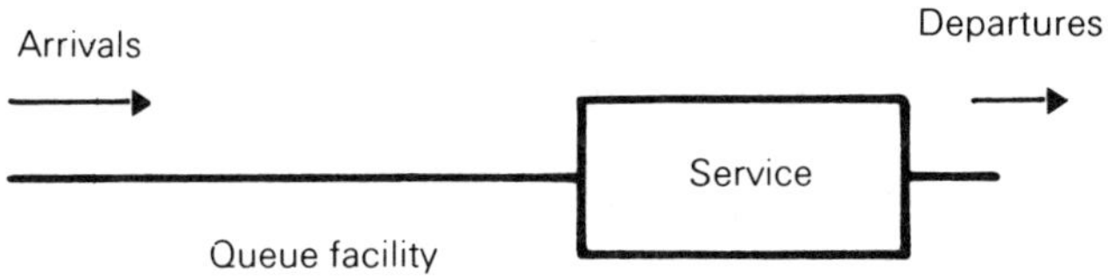

Figure 5.1

2. The queue (or waiting-line), which may be of zero length and for which the queue discipline gives a description of how items leave the queue in entering the service.
3. The service, which processes items according to some service-time distribution and which may consist of more than one service channel, in which case the multiplicity of channels operates in parallel. The number of items upon which the service can simultaneously operate is exactly equal to the number of service channels.

In all of the above we have discounted the possibility of several service points in sequence but are concentrating solely on the single-queueing system where items, having been through the service, are deemed to leave the system and are therefore no longer considered. A useful notation for the description of this type of queueing system is attributed to D.G. Kendall and is as follows:

A1/A2/n

where A1 is a letter denoting the arrival time distribution,
A2 is a letter denoting the service time distribution,
n is the number of service channels operating in parallel as part of a single service for the processing of items entering the system.

The letter M has been chosen to represent the negative exponential distribution which is commonly used in the simplest of analyses. This is a tribute to Markov, who did a lot of the early work on queues.

A.K. Erlang, who also made a great study of queueing theory, has been recognized for his work inasmuch as the letter E (variously subscripted as we shall see later) has been chosen to indicate distributions which are compounds of various negative exponential distributions. The letter D is used to represent the uniform (constant) distribution.

5.2 THE M/M/1 QUEUE WITH UNLIMITED LENGTH

We begin our analysis with the simplest of all queue models, the M/M/1 queue. According to the Kendall notation, this represents a queue system with a single channel and whose inter-arrival time and service-time distributions are approximated by the negative exponential probability distribution.

Before we commence the development of our model, it is helpful to know exactly what sort of information we desire concerning the behavior of the traffic system. Since arrivals, and hence services, are random, we must rely on expected values and probability terms. Therefore some likely points of information would be the following:

1. Expected number of units in the system.
2. Expected number of units in the queue.
3. Expected time in the system.
4. Expected time in the queue.
5. Probability of n units in the system.
6. Probability of n units in the queue.
7. Probability of more than n units in the system.
8. Probability of waiting time greater than τ.
9. Probability of service facility being idle.
10. Average time spent in system.
11. Average waiting time.
12. Probability of a queue existing.

For this information to be supplied the answer must first be found to the basic question: what is the probability of n units in the system? (Note that we regard the system to be both the service and the queue. The arrivals form an input and the output is of no concern to us.)

It may be reasonable to assume that this probability varies from time to time, until a steady-state is reached. We denote the probability of n units in the system at time t by $p_n(t)$. We now look at ways in which the state $p_n(t + \mathrm{d}t)$ can be reached from time t over a period of time so small that no more than one arrival is possible and no more than one service is possible.

If λ is the average arrival rate for unit time, then for time $\mathrm{d}t$ the average arrival rate will be $\lambda\mathrm{d}t$. The probability of no arrivals in this time is given by the Poisson formula:

$$\begin{aligned} P(\lambda\mathrm{d}t, 0) &= ((\lambda\mathrm{d}t)^0 \exp(-\lambda\mathrm{d}t))/0! \\ &= \exp(-\lambda\mathrm{d}t) = (1 - \lambda\mathrm{d}t) \end{aligned}$$

since $\lambda\mathrm{d}t$ is so small.

Since the only other possibility is one arrival, the probability of this must be $\lambda\mathrm{d}t$. The same approach is true, of course, for the service where the average service rate in per unit time is μ. There are four mutually

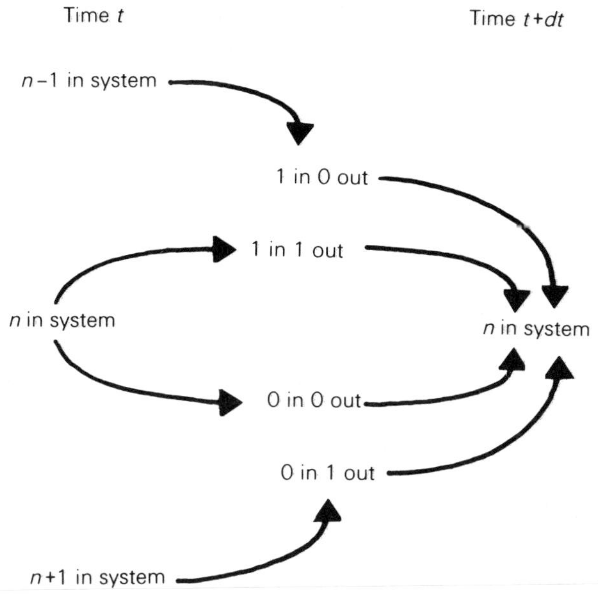

Figure 5.2

exclusive ways of approaching the state $p_n(t + \mathrm{d}t)$ from states at time t, as shown in Figure 5.2. Since these ways are mutually exclusive, their probabilities can be combined by simple addition, remembering that each is the simultaneous occurrence of three independent events. Therefore:

$$p_n(t + \mathrm{d}t) = p_{n-1}(t)\cdot\lambda \mathrm{d}t\,(1 - \mu \mathrm{d}t) + p_n(t)\cdot\lambda \mathrm{d}t\cdot\mu \mathrm{d}t + p_n(t)\cdot(1 - \lambda \mathrm{d}t)\cdot(1 - \mu \mathrm{d}t) + p_{n+1}(t)\cdot(1 - \lambda \mathrm{d}t)\cdot\mu \mathrm{d}t$$

This resolves to:

$$p_n(t + \mathrm{d}t) = \mathrm{d}t[\lambda p_{n-1}(t) + \mu p_{n+1}(t) - (\lambda + \mu)p_n(t)] + \mathrm{O}(\mathrm{d}t^2) + p_n(t)$$

Therefore

$$(p_n(t + \mathrm{d}t) - p_n(t))/\mathrm{d}t = \lambda p_{n-1}(t) + \mu p_{n+1}(t) - (\lambda + \mu)p_n(t)$$

Taking the limit as $\mathrm{d}t \to 0$, gives:

$$\mathrm{d}p_n(t)/\mathrm{d}t = \lambda p_{n-1}(t) + \mu p_{n+1}(t) - (\lambda + \mu)p_n(t) \tag{5.1}$$

It is now convenient to assume a steady-state when $p_n(t)$ is constant and therefore independent of time. Note that the phrase steady-state, as applied to a queue, does not refer to a state of unvarying behavior of the queue. The size of the queue will fluctuate in the steady-state but the statistical behavior of the queue is steady. On average, the proportion of an hour in which the queue length has some specific value will be the same from hour to hour in the steady-state. Likewise an item can expect, on average, to have to wait the same length of time whenever it joins a queue which is in the steady-state, although the waiting times of successive items will vary. Note also that if the average rates are not constant, the system will not reach a steady-state but it could remain stable. We shall see shortly that if the average arrival rate is not less than the average service rate, then the system is unstable and the probability of a long queue steadily increases.

Returning now to equation (5.1), and applying the condition for steady-state that the LHS vanishes, gives:

$$p_{n+1} = (1/\mu)\cdot[(\lambda + \mu)p_n - \lambda p_{n-1}] \tag{5.2}$$

This assumes $n > 0$, for if n were equal to zero we could not consider an item leaving at time t as shown in Figure 5.2, and p_{-1} is meaningless. Therefore the first subject of equation (5.2) will be p_2, but in fact we have not found p_1 and p_0 from previous analysis. We must now establish an equation involving only p_1 and p_0 and later determine p_0 (the probability that the system is idle).

There are only two ways in which the state $p_0(t + \mathrm{d}t)$ can be reached from states at time t. First, there is nothing in the system and nothing arrives (it is certain nothing will leave). Secondly, there is one in the system and it leaves and nothing arrives. Mathematically this is expressed by the following equation:

$$p_0(t + \mathrm{d}t) = p_0(t)\cdot(1 - \lambda\mathrm{d}t) + p_1(t)\cdot(1 - \lambda\mathrm{d}t)\cdot\mu\mathrm{d}t$$

therefore

$$p_0(t + \mathrm{d}t) = p_0(t) - \lambda p_0(t)\mathrm{d}t + p_1(t)\mu\mathrm{d}t + \mathrm{O}(\mathrm{d}t^2)$$

Taking the limit as $\mathrm{d}t \to 0$ gives:

$$\mathrm{d}p_0(t)/\mathrm{dt} = -\lambda p_0(t) + \mu p_1(t) \tag{5.3}$$

Again, in the steady-state, the derivative vanishes, giving:

$$p_1 = (\lambda/\mu)p_0 \tag{5.4}$$

We can now substitute for p_1 in equation (5.2) giving p_2 where now

$$p_2 = (1/\mu)\cdot[(\lambda + \mu)\cdot(\lambda/\mu)p_0] - \lambda p_0$$

or

$$p_2 = (\lambda/\mu)^2 p_0 \tag{5.5}$$

Successive applications of this equation give:

$$p_n = (\lambda/\mu)^n p_0 \qquad n > 0 \tag{5.6}$$

To find p_0 we simply apply our knowledge that a probability distribution, in order to be legal, must obey the rule that the sum of its probabilities must equal unity, i.e.

$$\Sigma_{n=0,\infty}\, p_n = 1 \tag{5.7}$$

In our case

$$p_0 + p_1 + p_2 + \ldots + \ldots = 1$$

or

$$p_n(1 + (\lambda/\mu) + (\lambda/\mu)^2 + (\lambda/\mu)^3 + \ldots) = 1$$

The terms in the brackets are those of an infinite geometric progression with first term equal to one and ratio λ/μ. Therefore:

$$p_0(1/(1 - (\lambda/\mu))) = 1 \tag{5.8}$$

or

$$p_0 = 1 - (\lambda/\mu)$$

This gives the probability of the service being idle, but of ultimate concern we get the final equation in equation (5.9) defining the steady-state probability of n units in the system:

$$p_n = (1 - (\lambda/\mu))\cdot(\lambda/\mu)^n \tag{5.9}$$

The ratio (λ/μ) is known as the traffic intensity and is usually denoted by ϱ. Clearly the analysis assumes $\varrho < 1$, or in other words the average arrival rate λ must be less than μ, the average service rate, although this does not prevent the build-up of a queue. It does however mean that the queue will not build up without limit.

Using equation (5.9) we are now in a position to evaluate some useful statistics:

1. The average number in the system is the expectation of the random variable n given by the probability distribution p_n. This parameter, usually denoted by N, is found from:

$$\begin{aligned} N &= \Sigma_{n=0,\infty}\, np_n && (5.10)\\ &= 0 + p_1 + 2p_2 + 3p_3 + \ldots \\ &= (1 - \varrho)\cdot(\varrho + 2\varrho^2 + 3\varrho^3 + \ldots) \\ &= (1 - \varrho)\cdot\varrho J && (5.11) \end{aligned}$$

where

$$J = 1 + 2\varrho + 3\varrho^2 + \ldots \tag{5.12}$$

Now

$$\int J \mathrm{d}\varrho = \varrho + \varrho^2 + \varrho^3 + \ldots = \varrho/(1 - \varrho)$$

therefore

$$J = \frac{\mathrm{d}[\varrho/(1 - \varrho)]}{\mathrm{d}\varrho} = 1/(1 - \varrho)^2 \tag{5.13}$$

Hence

$$N = \varrho/(1 - \varrho) \tag{5.14}$$

2. What is the expected number in the service? Well, it is zero when the system is idle, and one when the system is not idle. Therefore expected number in the service, denoted by u, is given by

$$u = 0 \cdot p_0 + (1 - p_0) = \varrho \tag{5.15}$$

3. Similarly the expected number in queue, usually denoted by Q, is the expected number in system minus the expected number in the service, i.e.

$$Q = \varrho/(1 - \varrho) - \varrho = \varrho^2/(1 - \varrho) \tag{5.16}$$

The above three statistics give an elementary spatial description of the M/M/1 queue system but we look now to a temporal one and we begin by forming the probability distribution of waiting time. An item will wait depending on the length of queue, i.e. on p_n. Suppose an item arrives with n items in the system, i.e. one in the service and $n - 1$ in the queue. First of all the queue must be reduced to zero and secondly the item in the service which will be its neighbor as it joins must have its service completed before the arrival ceases to wait. We want the arrival to wait time t and no longer than $t + \mathrm{d}t$, in which case the queue must 'lose' $(n - 1)$ items in a time t and the last item is served within a time t to $t + \mathrm{d}t$. This all depends on n. Items leave the queue at an approximately constant average rate μ. Hence the probability of $n - 1$ items leaving in time t is given by the Poisson distribution with mean value μt:

$$P(\mu t, n - 1) = ((\mu t)^{n-1} \exp(-\mu t))/(n - 1)!$$

The probability of a service in time $\mathrm{d}t$ is $\mu \mathrm{d}t$, so that a waiting time distribution is given by:

$$w(t) = \Sigma_{n=0,\infty}\, p_n (\mu t)^{n-1} \exp(-\mu t)/(n - 1)!$$

(Note that the items do not wait if $n = 0$ upon arrival.)
Therefore

$$\begin{aligned} w(t)\mathrm{d}t &= \Sigma_{n=1,\infty}\ ((1-\varrho)(\varrho\mu)^n)/(n-1)!\ t^{n-1}\exp(-\mu t)\mathrm{d}t \\ &= (1-\varrho)\exp(-\mu t)\mathrm{d}t\ \{(\varrho\mu) + (\varrho\mu)^2 t + (\varrho\mu)^3 t^2/2! + \ldots\} \\ &= (\varrho\mu)\cdot(1-\varrho)\cdot\exp(-\mu t)\mathrm{d}t\cdot\{1 + (\varrho\mu t) + (\varrho\mu t)^2/2! + \ldots\} \\ &= (\varrho\mu)\cdot(1-\varrho)\cdot\exp(-\mu t)\mathrm{d}t\cdot\exp(\varrho\mu t) \\ &= (\varrho\mu)\cdot(1-\varrho)\cdot\exp(-(1-\varrho)\mu t)\mathrm{d}t \end{aligned}$$

Therefore

$$w(t) = (\varrho\mu)\cdot(1-\varrho)\cdot\exp(-(1-\varrho)\mu t)$$

or

$$w(t) = (\lambda/\mu)(\mu-\lambda)\exp(-(\mu-\lambda)t) \tag{5.17}$$

We may now continue to obtain further statistics giving us our temporal description of the M/M/1-queue:

4. The average waiting time t_w is given by:

$$t_w = \int t w(t)\mathrm{d}t = \varrho/\mu(1-\varrho) \qquad \text{(integrating from 0 to } \infty) \tag{5.18}$$

$$t_w = N/\mu = Q\lambda \tag{5.19}$$

5. The probability of a waiting time greater than τ is:

$$\begin{aligned} w(t > \tau) &= \int w(t)\mathrm{d}t \qquad \text{(integrating from } \tau \text{ to } \infty) \\ &= (\lambda/\mu)\exp(-(\mu-\lambda)t) \end{aligned} \tag{5.20}$$

6. The probability distribution of time spent in the system can be similarly derived. In this case the n items in the system found upon arrival by an item entering the system must be dealt with within a time t and the newly arrived item must then be served within a further period of time $\mathrm{d}t$. The combined probabilities of these events, summed for all possible values of n, gives the system time distribution as follows:

   $$S\mathrm{d}t = \Sigma_{n=0,\infty}\ p_n\ P(\mu t,n)\mu\mathrm{d}t$$

   where

   $$P(\mu t, n) = (\mu t)\ \exp(-\mu t)/n!$$

   Therefore

   $$\begin{aligned} S(t)\mathrm{d}t &= \mu(1-\varrho)\ \exp(-\mu t)\ \Sigma_{n=0,\infty}\ (\varrho\mu t)^n/n!\ \mathrm{d}t \\ &= \varrho(1-\varrho)\cdot\exp(-(1-\varrho)\mu t)\mathrm{d}t \end{aligned}$$

$$S(t) = (\mu - \lambda)\exp(-(\mu - \lambda)t) \qquad (5.21)$$

7. The probability of spending longer than a time τ in the system is:

$$S(t > \tau) = \int (\mu - \lambda)\exp(-(\mu - \lambda)t)\mathrm{d}t \qquad \text{(integrating from } \tau \text{ to } \infty)$$
$$= \exp(-(\mu - \lambda)\tau) \qquad (5.22)$$

Notice that the probability of spending less than a time τ in the system is:

$$S(t < \tau) = \int (\mu - \lambda)\exp(-(\mu - \lambda)t)\mathrm{d}t \qquad \text{(integrating from 0 to } \tau)$$
$$= 1 - \exp(-(\mu - \lambda)\tau) \qquad (5.23)$$

and that the sum of this and the probability given by equation (5.22) is unity, i.e. it is certain that items entering the system will spend some time in it.

8. However, the probability of a waiting time less than τ is:

$$w(t < \tau) = \int w(t)\mathrm{d}t \qquad \text{(integrating from 0 to } \tau)$$
$$= (\lambda/\mu)(1 - \exp(-(\mu - \lambda)\tau)) \qquad (5.24)$$

and the sum of this and the probability given by equation (5.20) is not unity but is equal to (λ/μ), i.e. the probability of having to wait is (λ/μ).

The probability of not having to wait is therefore $1 - (\lambda/\mu)$, which is therefore the probability of the system being idle, confirmed, of course, by the value of p_0, equation (5.8). The utilization of the system is the probability of the system not being idle and for the M/M/1 queue this is identical to the traffic intensity ϱ. The following statistics complete a fairly detailed description, both spatial and temporal, of the M/M/1 queueing system:

9. The average time spent in the service is the reciprocal of the average service rate $= 1/\mu$.
10. The average time spent in the system, t_s, can be found from either:

    (a) the sum of the average waiting time t_w plus the average time spent in the service, i.e.

    $$t_s = t_w + 1/\mu = 1/(\mu - \lambda) \qquad (5.25)$$

    (b) the definition of the expectation of the random variable, time spent in the system, i.e.

    $$\int tS(t)\mathrm{d}t \qquad \text{(integrating from 0 to } \infty)$$
    $$= \int t(\mu - \lambda)\exp(-(\mu - \lambda)t)\mathrm{d}t = 1/(\mu - \lambda)$$

as expected.

11. The probability that a queue exists is identically the probability of more than one in the system, i.e.

$$\Sigma_{n=2,\infty}\, p_n = 1 - (p_0 + p_1) = \varrho^2$$

12. Finally, the probability that there will be more than k items in the system is given by:

$$p_a > k = \Sigma_{n=k+1,\infty}\, p_n = \varrho^{k+1} \qquad (5.26)$$

Before other types of queue system are considered, the reader is advised to work through the following examples, all of which relate to the M/M/1 queue.

Example 5.1

Aircraft arrivals at a particular airport demonstrate an arrival-time distribution that is negative exponential with a mean of 2 minutes. A procedural controller services these arrivals with a negative exponential service-time distribution at a mean rate of 40 aircraft per hour. This is done on a first-come, first-served basis, waiting aircraft having to fly a holding pattern whilst the controller handles items on an individual basis before passing them on to his no. 1 director. The system constituted by the controller and the aircraft stack is satisfactorily represented by an M/M/1 queue.

The following data are required:

1. The probability that an aircraft will have to wait in a stack before the controller can direct it on to his no. 1 director.
2. The average number of aircraft in the system comprising those in the stack and under the direction of the procedural controller.
3. The amount of time aircraft can expect to spend in this system.
4. It is thought that an additional procedural controller would be needed if the aircraft's average time in this system exceeds 10 minutes. How much must the average rate of arrivals increase to warrant a second controller?

First we recognize that the average arrival rate, λ, is the reciprocal of the mean arrival time, τ, i.e. $\lambda = 1/\tau$, and $\tau = 2$ min, therefore $\lambda = 30$/h. Also the value of μ is given, i.e. $\mu = 40$/h. The traffic intensity, ϱ, is therefore 0.75.

1. The probability of having to wait is identically the probability of non-zero items in the system, i.e. $1 - p_0$. Since $p_0 = 1 - \varrho$, this probability is identically the traffic intensity, i.e. 0.75.

2. The average number of aircraft in the system is given by equation (5.14), therefore $N = \varrho/(1 - \varrho) = 3$.
3. The average time spent in the system is given by equation (5.25), therefore $t_s = 1/(40 - 30) = 6\,\text{min}$.
4. If the value of t_s exceeds 10 min, then a second controller would be needed. Keeping μ at its value of 40/h, this gives $60/(40 - \lambda) = 10$ or $\lambda = 34$. Therefore arrivals would need to increase by more than four per hour on average in order to warrant a second controller.

Example 5.2

The demand for international telephone calls made in a certain gateway exchange is given by a Poisson distribution with an average rate of one per minute. The exchange equipment has a negative exponential service-time distribution with an average of 20 seconds to route calls. Assume that calls are handled individually, on a first-come, first-served basis, and that if the exchange is busy when further calls arrive, these must await processing in a queue.

1. What is the average time that a call spends waiting in the system?
2. What proportion of calls will wait less than half a minute?
3. What is the average number of calls stored in the queue?
4. If the size of the queue is ten, by how much will the traffic need to build up before it becomes necessary to install faster switching equipment?

Once again we first determine values for λ and μ from the data given, i.e. $\lambda = 1/\text{min}$, $\mu = 3/\text{min}$ ... $\varrho = 1/3$.

1. The average time spent waiting is given by equation (5.18), i.e.

$$t_w = \varrho/\mu(1 - \varrho) = (1/3)(3*2/3) = 1/6 = 10\,\text{s}$$

2. The proportion of calls waiting for less than half a minute is given by equation (5.24):

$$w(t < \tau) = (\lambda/\mu)(1 - \exp(-(\mu - \lambda)\tau)$$
$$w(t < 0.5) = (1/3)\cdot[1 - \exp(-2*0.5)] = 0.2107$$

Therefore about 1/5 of calls will wait less than 30 s.

3. The average number of calls stored in the queue is given by equation (5.16):

$$Q = \varrho^2/(1 - \varrho) = (1/9)(2/3) = 1/6$$

4. It will be necessary to install faster switching equipment if, keeping μ at its value of 3/min until such time, the traffic intensity is such as to make

$$p^2/(1 - \varrho) > 10$$

or

$$\varrho^2 + 10\varrho - 10 > 0 \qquad \text{i.e. } \varrho = 0.916$$

Therefore if the average arrival rate exceeds 2.75 (approximately), faster equipment will be needed.

Example 5.3

A digital computer processing conflict search requests from the no. 1 director in an ATC system receives an average of 8 jobs/h. Many are short jobs, while others are quite long. Assume, however, that the time to perform a job has an exponential distribution with a mean of 6 minutes.

1. What is the average elapsed time from the time a request is made until it is completed by the computer?
2. What is the average number of jobs in the system waiting service and being processed?
3. What is the likelihood of there being more than five jobs in the system?
4. If the computer is deemed to be overloaded if a job has to spend more than 25 minutes in the system, 80 percent of the time, by how much would the arrival rate need to increase before overloading of this nature occurred?

1. The average time spent by jobs in the system is given by equation (5.25). Noting that the values of λ and μ are 8 and 10 respectively, taking 1 hour as unit time, then

$$t_s = 1/(\mu - \lambda) = 30\,\text{min}$$

2. The average number of jobs in the system is given by equation (5.14), therefore

$$N = \varrho/(1 - \varrho) = (4/5)(1 - 4/5) = 4$$

3. The probability of there being more than any given number of jobs in the system is given by equation (5.26), therefore

$$p_n > 5 = \varrho^6 = 0.8^6 = 0.262$$

4. The probability that a job has to spend more than a stated amount of time in the system is given by equation (5.22). If this probability is 0.8 for a value of time equal to 25 min, or 5/12 h, then this gives us

$$0.8 = \exp(-(\mu - \lambda)*5/12)$$

where $\mu = 10$ and λ is then determined from the above equation. Taking natural logarithms gives

$$-0.22314355 = -(10 - \lambda)\cdot 5/12$$

whence $\lambda = 9.46$. Therefore if the average arrival rate exceeds 9.46 jobs/hour then the computer will be deemed overloaded. This is equivalent to an increase of more than 18 percent.

5.3 THE M/M/1 QUEUE WITH LIMITED LENGTH

We now turn to the analysis of more complex queue problems and the first step to take is to consider again the M/M/1 queue but with the added constraint of a limited length of the queue. This is quite obviously a physical constraint for many traffic situations and it is important therefore that we should be able to determine some measures of the system state with this condition applied. Suppose that the limit on the length of the queue is L. Any items arriving when the size of the queue is L do not join the queue. In this case, as the number in the system is limited, a steady-state is possible for any value of ϱ and not only when the average arrival rate is less than the average service rate. However, if L is large and ϱ is greater than one, the steady-state may be reached very slowly and in practice may not be achieved before the operating conditions change. Our analysis proceeds as follows. The arrival rate is more strictly described as joining-the-queue rate and so for $n < L$ it is one, as before, but for $n > L$ it is zero. The service rate is unaffected by the limiting condition on queue length. Hence

$$p_n = \varrho^n p_0 \qquad \text{for } n < L \tag{5.27}$$

and

$$p_n = 0 \qquad \text{for } n > L$$

The value of p_0 is changed however, since, now:

$$\begin{aligned}\Sigma_{k=0,L}\, p_n = 1 &= p_0\, \Sigma_{k=0,L}\, \varrho^k \\ &= p_0\,(1 - \varrho^{L+1})(1 - \varrho)\end{aligned}$$

Theretore

$$p_0 = (1 - \varrho)/(1 - \varrho^{L+1}) \tag{5.28}$$

This, of course, is also the idle time of the system and the utilization is still $1 - p_0$, therefore

$$u = \varrho(1 - \varrho^L)/(1 - \varrho^{L+1}) \tag{5.29}$$

By definition, the expected number in the service is the probability of the service being busy, i.e. its utilization, as before. Although substituting $\varrho = 1$ into equation (5.28) gives an indeterminate result, a value for p_0 can be found for $\varrho = 1$ from equation (5.27), i.e.

$$\Sigma_{k=0,L}\, p_k = 1 = p_0\, \Sigma_{k=0,L}\, 1^k = p_0(L + 1)$$

i.e.

$$p_0 = 1/(L + 1) \qquad (\text{when } \varrho = 1) \tag{5.30}$$

We now proceed to determine the following measures of system state:

1. The proportion of items lost to the system due to the limited length of queue.
2. The average number of items in the system.
3. The average number of items in the queue.

1. Items are lost on account of the system being full when they arrive. The probability of this is of course p_L. Therefore the proportion of such items is given by

$$p_L = \varrho^L(1 - \varrho)/(1 - \varrho^{L+1}) \tag{5.31}$$

2. By definition, the average number in the system, N, is given by:

$$\begin{aligned} N &= \Sigma_{k=0,L}\, kp_k \qquad (5.32) \\ &= 0 + p_1 + 2p_2 + 3p_3 + 4p_4 + \ldots + Lp_L \\ &= p_0\,\{\varrho + 2\varrho^2 + 3\varrho^3 + \ldots L\varrho^L\} \\ &= \varrho p_0\,\{1 + 2\varrho + 3\varrho^2 + \ldots + L\varrho^{L-1}\}\end{aligned}$$

The series in the brackets computes to:

$$((1 - L\varrho^L) - \varrho^L(1 - L\varrho))/(1 - \varrho)^2 \quad (5.33)$$

Whence

$$N = (\varrho - (L + 1)\varrho^{L+1} + L\varrho^{L+2})/((1 - \varrho)(1 - \varrho^{L+1})) \quad (5.34)$$

3. The expected number in the queue is the expected number in the system minus the expected number in the service. Therefore:

$$Q = N - u = N + p_0 - 1 \quad (5.35)$$

$$Q = (\varrho - (L + 1)\varrho^{L+1} + L\varrho^{L+2})/((1 - \varrho)(1 - \varrho^{L+1}))$$
$$= \varrho[(1 - (L + 1)\varrho^L + L\varrho^{L+1})/((1 - \varrho)(1 - \varrho^{L+1}))$$

5.4 MULTIPLE SERVICE CHANNELS

Our next consideration is that of multiple service channels. Here we now see a number of service channels operating independently in parallel, but from the systems point of view constituting a single service. Nevertheless, if there are, say, c such channels, then the number of items which may be served simultaneously is c. The c service channels each have the same average service rate, μ, so that from the systems' viewpoint the effective service rate is $c\mu$ when a queue has formed, otherwise it is $n\mu$ when $n < c$. So, λ is as before and $\mu = n\mu$ for $n < c$ and $\mu = c\mu$ for $n > c$.

We must again develop an expression for p_n, the probability that in the steady-state there are n items in the system. We proceed as follows:

1. The probability of a service in time dt in one channel $= \mu dt$.
2. The probability of no service in time dt in one channel $= 1 - \mu dt$.
3. The probability of no service in time dt in n channel $= (1 - \mu dt)^n$.
4. Therefore the probability of at least one service being completed in a time dt when n service channels are occupied is

   $$1 - (1 - \mu dt)^n = n\mu dt$$

   ignoring dt^2 terms and those of higher order.
5. Finally, the probability of just one service being completed in a time dt when n service channels are occupied is the probability in 4 multiplied by probability that the remaining $(n - 1)$ do not go

   $$= n\mu(1 - \mu dt)^{n-1}dt$$

Proceeding as with the M/M/1 queue we get:

$$p_n(t + dt) = p_n(t)(1 - \lambda dt)(1 - \mu dt)^{n-1}$$

$$+ p_n(t)\lambda dt(n\mu dt)(1 - \mu dt)^{n-1}$$
$$+ p_{n+1}(t)(1 - \lambda dt)(n + 1)\mu dt(1 - \mu dt)^n$$
$$+ p_{n-1}(t)\lambda dt(1 - \mu dt)^{n-1} \quad (5.36)$$

This reduces to:

$$dp_n(t)/dt = \mu(n + 1)p_{n+1}(t) - (\lambda + n\mu)p_n(t) + \lambda p_{n-1}(t) \quad (5.37)$$

In the steady-state, this gives:

$$p_{n+1} = 1/(\mu(n + 1))\cdot(\lambda + n\mu)p_n + \lambda p_{n-1} \quad (5.38)$$

This is true for $n > 0$, but for $n = 0$ we get:

$$p_0(t + dt) = p_0(t)(1 - \lambda dt) + p_1(t)(1 - \lambda dt)\mu dt$$

as with the M/M/1 queue, therefore:

$$p_1 = (\lambda/\mu)p_0$$

Repeated application of equation (5.38) gives:

$$p_n = (\lambda/\mu)^n p_0/n! \quad (5.39)$$

This equation holds for $n < c$ but thereafter queues form and the queue leaving rate is determined by all services being full:

$$p_n(t + dt) = p_n(t)\cdot(1 - \lambda dt)\cdot(1 - \mu dt)^c$$
$$+ p_{n+1}(t)\cdot(1 - \lambda dt)c\mu dt(1 - \mu dt)^{c-1}$$
$$+ p_{n-1}(t)\cdot\lambda dt\cdot(1 - \mu dt)^c \quad (5.40)$$

(The case for $p_n(t)$ with one in and one out is omitted since this always contributes a dt^2 term.) Equation (5.40) reduces to:

$$dp_n(t)/dt = \mu c p_{n+1}(t) - (\lambda + c\mu)p_n(t) + \lambda p_{n-1}(t) \quad (5.41)$$

And, in the steady-state, gives:

$$p_{n+1} = (1/c\mu)\cdot(\lambda + c\mu)p_n + \lambda p_{n-1} \quad (5.42)$$

for $n > c$.

Now, since

$$p_c = (\lambda/\mu)^c p_0/c!$$

then

$$p_{c+1} = 1/c - (\lambda/\mu)^{c+1} p_0/c! \quad (5.43)$$

and so

$$p_{c+j} = 1/c^j - (\lambda/\mu)^{c+j} p_0/c! \quad (5.44)$$

letting $c + j = n$, and noting that $n > c$ then

$$p_n = c^{c-n} \varrho^n p_0/c! \qquad (5.45)$$

Equations (5.39) and (5.45) summarize the steady-state probabilities for the M/M/c queue, for which the probability that there are n items in the system is given by the formulae:

$$p_n = \varrho^n p_0/n! \qquad \text{for } n < c$$

and

$$= c^{c-n} \varrho^n p_0/c! \qquad \text{for } n > c$$

Using the definition that

$$\Sigma_{n=0,\infty}\ p_n = 1$$

it is simple to show that p_0 is given by equation (5.46):

$$p_0 = \{(\Sigma_{n=0,c-1}\ \varrho^n/n!) + (\varrho^c/((c-1!)(c-\varrho n)))\}^{-1} \qquad (5.46)$$

An expression for the average number of items in the system may also be derived from the basic definition and this is:

$$N = (\varrho^{c+1}\ p_0/(c-\varrho)^2(c-1)!)) - \varrho \qquad (5.47)$$

For completeness, other measures of state for the system M/M/c are as follows:

1. Utilization,

$$u = \varrho/c. \qquad (5.48)$$

2. $$P_{n>k} = c^{c-k} \varrho^{k+1} p_0/(c-\varrho)c! \qquad (5.49)$$

3. $$Q = N - \varrho = \varrho^{c+1} p_0/(c-\varrho)^2(c-1)! \qquad (5.50)$$

4. The probability of waiting for greater than a time τ is

$$w(t > \tau) = p_{n>c-1} \cdot \exp - \mu\tau(c - \varrho) \qquad (5.51)$$

5. The average waiting time,

$$t_w = Q/\lambda \qquad (5.52)$$

6. Finally, the average time spent in the system, t_s, is

$$t_s = t_w + 1/\mu = N/\lambda \qquad (5.53)$$

The measures of state found for the M/M/1 and M/M/c queues will enable us now to make a comparison of the two by means of an elementary example. Suppose that at a self-service store there is a single cashier. We further assume that arrival and service times are random and therefore supply us with the basis of an M/M/1 queue. Suppose that on average nine customers arrive every five minutes and that the cashier can serve ten customers in a period of five minutes, on average. This gives us $\lambda = 1.8$, $\mu = 2.0$ and $\varrho = 0.9$ (measuring time in minutes).

The features of the system which may be important to the store manager are as follows:

1. The average number of customers waiting for service (which may decide whether a shopper comes into the store), which is

   $$Q = \varrho^2/(1 - \varrho) = 0.81/0.1 = 8.1$$

2. The probability of having more people waiting than there is convenient room for, say ten altogether

   $$p_{n>10} = \varrho^{11} = 0.9^{11} \times 0.31$$

3. The probability of a customer having to wait too long for service, say more than two minutes

   $$w(t > 2) = \varrho \exp(-\ 2(\mu - \lambda)) = 0.6$$

   By comparison, if service can be speeded up to, say, 12 in 15 minutes by, for example, using a superior cash register, then the quantities above become $\lambda = 1.8$, $\mu = 2.0$ and $\varrho = 0.75$; $Q = 2.25$, $p_{n>10} = 0.042$, $w(t > 2) = 0.23$.

   A slight change in the service rate, in this case an increase of 20 percent, makes a significant difference in the operation of the system.

Let us now consider the situation for a second cashier. Suppose as before that $\mu = 2$ for each cashier and $\lambda = 1.8$. Then $\varrho = 0.9$ and $c = 2$. Then we calculate the following:

1. Using equation (5.46): $p_0 = 0.38$.
2. Using equations (5.47) and (5.50): $Q = 0.23$.
3. Using equation (5.49): $P_{n>10} = 0.00021$.
4. Using equation (5.51): $w(t > 2) = 0.0034$.

Furthermore, the probability of having to wait at all is given by:

$$p_{n>1} = 0.28.$$

The introduction of a second service channel has therefore reduced the queueing to a very low level. It is of interest to notice what would happen if the arrival rate for the M/M/2 queue increased so as to restore the original value for system utilization, i.e. $\lambda = 3.6$, $\mu = 2.0$, $c = 2$ and

$u(=\varrho/c) = 0.9$ as for the M/M/1 queue with $\lambda = 1.8$. With these figures we find the following: $Q = 7.7$, $p_{n>10} = 0.33$ and $w(t > 2) = 0.38$. Thus although the number of customers is doubled, the numbers waiting are about the same and the probability of a long wait is a lot less. The system is thus not equivalent to the sum of two separate stores each like the original M/M/1 example.

This analysis illustrates three features which are common to most queueing situations:

1. Unless utilization of the service is fairly high (over 0.5 say), the congestion is very slight.
2. When the utilization approaches unity, a small change in it makes a large difference to the congestion.
3. When service channels can be shared by one queue, the congestion is much less than it would be for the same rates of input and service but with an independent queue for each service channel.

5.5 ERLANG QUEUES

We come to our final consideration in the analysis of queues and in this we examine the application of a probability distribution function hitherto mentioned but not discussed in any fine detail, namely the Erlang distribution. With the negative exponential distribution applied to both arrival and service times, it is assumed, respectively, that the probability of an arrival at any instant is quite independent of previous arrivals, and the probability of a service being completed at any instant is quite independent of the time when that service began. It is this property of independence which allows simple transition equations to be written. In practice the unrelated arrival of customers might well be negative exponential, but the distribution of service times may be quite unlike the negative exponential.

An appreciable deterioration in service times is usual with the passage of time and when it cannot be assumed that the service rate is independent of when the service began, an alternative approach must be found before the transition equations can be developed satisfactorily. One method is to assume that a service-time distribution is built up of several exponential distributions. In this way, the service is considered as a series of phases immediately following each other and each possessing an exponential distribution. The distribution of the whole service time is quite different from that of its components. When all phases, say a total of k, are identical, the whole service time has a chi-square distribution (refer to equation (5.54)) with $2k$ degrees of freedom. Such combinations of exponential distributions have come to be known as Erlang distributions, in tribute to A.K. Erlang, from whose work this field of study derived much benefit.

The basic method is to represent the distribution in question by a number of routes, along one of which the item requiring service passes according to an assigned probability distribution, and each route is assembled from exponential phases. The transition equations for systems with such distributions of service times can now be written, because the probability of passing from the jth to the $(j + 1)$th phase of the service is $mj\mathrm{d}t$, and it depends only on the presence of an item in the jth phase. For random arrivals and Erlang service time, a single-channel system is characterized by $p(n, j)$ which is the probability of n units waiting and the item being served as in the jth phase of service. If we assume an overall service rate of m and k identical phases, then the service rate for each phase is km, and if l is the arrival rate, then the following transition equations hold:

$$
\begin{aligned}
p(0,0) &= p(0,0)(1 - l\mathrm{d}t) + p(1, k)(1 - l\mathrm{d}t)km\mathrm{d}t \\
p(1,1) &= p(1,1)(1 - l\mathrm{d}t)(1 - km\mathrm{d}t) + p(2,k)(1 - l\mathrm{d}t)km\mathrm{d}t \\
&\quad + p(0,0)l\mathrm{d}t \\
p(n,1) &= p(n,1)(1 - l\mathrm{d}t)(1 - km\mathrm{d}t) + p(n + 1,k)(1 - l\mathrm{d}t)km\mathrm{d}t \\
&\quad + p(n - 1,1)l\mathrm{d}t(1 - km\mathrm{d}t) \\
p(1,j) &= p(1,j)(1 - l\mathrm{d}t)(1 - km\mathrm{d}t) + p(1,j - 1)(1 - l\mathrm{d}t)km\mathrm{d}t \\
p(n,j) &= p(n,j)(1 - l\mathrm{d}t)(1 - km\mathrm{d}t) + p(n - 1,j)l\mathrm{d}t(1 - km\mathrm{d}t) \\
&\quad + p(n,j - 1)(1 - l\mathrm{d}t)\mathrm{kmd}t
\end{aligned}
$$

In general, the solution of these equations requires the introduction of probability generating functions and numerical computation. For $k = 2$, it can be shown that

$$p(n) = \Sigma_{i+0,n} \{(n + i + 1)!(1 - \varrho)/(2i + 1)!(n - i)!\}\cdot(\varrho/2)^{n+1} \quad (5.54)$$

where $\varrho = \lambda/\mu$, but for larger values of k, the probabilities have to be found numerically.

The customary statistics are given for the $M/E_K/1$ queue system as follows:

1. Service-time distribution

$$m(t) = (\mu k)^k \exp(-\mu kt)t^{k-1}/(k - 1)! \quad (5.55)$$

2. Utilization,

$$u = \varrho. \tag{5.56}$$

3. Probability of idle system,

$$p_0 = 1 - \varrho. \tag{5.57}$$

4. Average number in system,

$$N = Q + \varrho. \tag{5.58}$$

5. Average number in queue,

$$Q = (k + 1)\varrho^2/(2k(1 - \varrho)). \tag{5.59}$$

6. Average waiting time,

$$t_w = Q/\lambda. \tag{5.60}$$

7. Average time spent in system,

$$t_s = N/\lambda. \tag{5.61}$$

When $k \to \infty$ the service time is constant at $1/\mu$ so that

$$Q = 0.5*\varrho^2/(1 - \varrho) \tag{5.62}$$

Example 5.4

A digital computer system has a large number of printed circuit boards which break down at random intervals. On average one fails every eight months. A computer systems engineer carries out a detailed maintenance schedule on each faulty board. The schedule consists of five distinct phases in sequence each taking, on average, approximately two weeks, but the actual time taken varies widely. How much time will the engineer have for other tasks (which are interrupted to deal with these repairs)? What is the average time that a board is out of use?

As there are a large number of boards, the rate at which they fail may be assumed independent of the number which have already failed. Thus the failure distribution may be taken to be negative exponential. Boards which fail are then serviced by the engineer and it is possible for a 'queue' of failed boards to form.

The service-time distribution, taken by the engineer, is not known, but each phase of the service may be assumed to be negative exponential so that total service time is Erlang with five phases and each having an average service rate of 1/2 taking unit time as 1 week. This gives the $M/E_5/1$ queue with $\lambda = 1/35$ and $\mu = 1/10$. Therefore:

$$\varrho = 10/35 = 2/7$$

The proportion of idle time, p_0 is given by equation (5.57), i.e.

$$p_0 = 1 - \varrho = 5/7$$

Therefore the engineer will be 'idle' for about 70 percent of his time and he will therefore spend this amount of his time on other tasks.

The average amount of time which a board is out of use is given by t_s, the average amount of time spent in the 'system'. Using equations (5.62), (5.58) and (5.55) gives:

$$Q = (5 + 1)(2/7)^2/(2*5*(1 - 2/7)) = 24/350$$
$$N = 24/350 + 2/7 = 124/350$$
$$t_s = (124/350)*35 = 12.4 \text{ weeks}$$

Had we assumed the service to be of constant service time, giving the M/D/1 queue, then the value of t_s is found as follows, using firstly equation (5.62):

$$Q = (2/7)^2/(2*(1 - 2/7)) = 4/49*7/10 = 4/10$$
$$N = 4/70 + 2/7 = 24/70$$
$$t_s = (24/70)*35 = 12 \text{ weeks}$$

5.6 CONCLUSIONS

In this chapter on the very important subject of waiting-line design, we have analyzed simple queueing systems with Poisson arrivals and service-time distributions of two types: negative exponential and Erlang. With little difficulty more arbitrary service times such as constant service time could be handled. All our systems had an infinite source. However, here again, it creates no serious problems to analyze situations where the number of arrivals is finite. The consequence of this is that the effective arrival rate is a function of the service rate of each server (multiple-server system), the number of servers, the number of 'customers' in the calling population and the distribution of 'the time until service is needed' for each customer.

In addition, the systems we have examined had: limited waiting-lines, unlimited waiting-lines, single server and multiple servers. There are other systems that can be analyzed analytically; however, the system need not be very complex before analytical solutions are no longer available, in which case the customary recourse is to simulation. For example, with a system having many servers and many waiting-lines, customers may do one of the following:

1. See that all lines are too long, so they leave without being served.
2. Jump from one line to another.
3. Arrive according to a non-Poisson process.
4. Be given a certain priority.

In addition one or more servers may be in the following states:

1. Be servicing at different rates depending on the type of customer.
2. Shut down for a break or breakdown.
3. Be interrupted by maintenance, a supervisor, etc.
4. Service faster as the queue becomes longer.

In these situations, the only possible means of analysis may be by simulation, and of specific interest to analysts are the digital computer simulation languages available.

Most realistic simulation projects are such that a separate computer program is required for each project. Consequently, one of the main objectives is to get a fairly efficient executable program running in a short period of time. Obviously the program should not be written in machine language or assembly language, but what about a general-purpose language such as Fortran or PL/1? Both of these are very general in nature and can be used for any simulation project. In fact, Fortran is used quite often because it is well known, available on almost all computer systems and generally efficient in time as well as in storage. The general nature of Fortran often precludes its use for the special purpose realistic projects requiring simulation such as, for example, the analysis of traffic systems. It is true that PL/1 is more sophisticated since it possesses Fortran capabilities within it but additionally has list-processing facilities and many others quite well suited to the intricate 'housekeeping' nature of many simulation projects.

There are a number of simulation languages such as Gasp and Simscript, purpose-built almost to suit the particular type of simulation needed. The most generally used simulation language is GPSS (general-purpose simulation system) which was developed by IBM. This is a problem-oriented language, but it has a wide range of applications. It is a variable time increment language which carries out simulations in integral time units. The GPSS user writes down the general flow of what is to be simulated in block diagram form and then blocks are written as GPSS statements. GPSS automatically performs 'book-keeping' chores, and also calculates and prints many important statistics. Once the language is learned, a program to simulate complex projects can be written in a very short period of time. Some of the many other simulation languages which have been developed are Dynamo, CSMP, Simpac, Simulate, GSP, Simula, ESP and CSL. As with any high-level language, the use of a simulation language generally decreases programming time but increases the computer running time and memory requirements. Consequently, careful judgment must be made in selecting the choice of simulation language over, say, Fortran or PL/1.

Finally, it should be realized that the complexity of the real-world situation that exists in traffic systems of all kinds is such that the analytical approach rapidly falls down, leaving simulation as the only way forward. Analytical techniques, as we have seen, provide useful insight into the behavior of traffic systems. However, the use of simulation coupled with heuristic techniques, often proves to be immensely more powerful in the case of highly stochastic and dynamic systems. In this regard, the work of Mars is of particular significance. The reader is encouraged to consult the references, to discover the power of heuristic techniques for simulating the performance of complex telecommunication systems. Learning automata offer an exciting paradigm for the design of algorithms which increase the efficiency of these systems in a contemporary real-world setting.

5.7 EXERCISES

1. A digital computer supervising an experiment receives Poisson distributed signal pulses of negligible duration at a mean rate of λ signals/s. The computer is capable of processing these signals at a mean rate of μ according to a negative exponential service-time distribution. Should signals arrive while previous signals are being processed then they are stored awaiting service, thereby forming a queue. At any chosen time, in the steady-state, the probability $p(n)$ of n signals in the system at that time, is:

$$p(n) = (\lambda \cdot p(n-1) + \mu \cdot p(n+1)/(\lambda + \mu)$$

(a) Show that the above equation may be simplified to:

$$p(n) = (\lambda/\mu)^n \cdot (1 - \lambda/\mu)$$

If $\lambda = 1$ and $\mu = 1.05$ then at any time calculate:

(b) the probability of the computer being idle,

(c) the expected number of signals in the system.

2. Aircraft arrivals at an airport are given by a Poisson probability distribution. Show that the inter-arrival times are given by a negative exponential distribution. If the mean arrival rate in unit time is λ, determine the average time between

arrivals. Arrivals are serviced by a single runway, μ being the mean service rate in unit time. If this is occupied, aircraft wishing to land form a queue by stacking. Derive an expression to find the probability of having n aircraft in the system. The mean arrival rate is 20/h and the average time to clear the runway is 2 min. Find the probability of there being any stacked aircraft. What is the expected number of these when the arrival rate is 25/h?

3. The probability of finding n items in a single channel queue system with Poisson arrivals and negative exponential service times is given by

$$\mathrm{d}p_n(t)/\mathrm{d}t = \lambda p_{n-1}(t) + \mu p_{n+1}(t) - (\lambda + \mu)p_n(t)$$

Derive an expression involving only n, λ and μ to find the steady-state probability p_n of having n items in the system. A certain process produces items in two stages. The output from the first stage proceeds to the input of the second stage where a queue facility is available.

The mean service rates of the first and second stages, based on Poisson arrival distributions and negative exponential service time are 10.1 and 10.4 items/hour respectively. Calculate the percentage idle time of the first stage if this shuts down when the number of items in the queue reaches four. Assuming now no limit to the size of the queue, determine the service rate for the first stage which minimizes the total cost. Cost of items in the queue is £3/item/hour, and idle time in the second stage costs £250/hour.

4. Starting from the binomial probability distribution, determine a probability model for calculating the number of telephone calls received at an exchange in a given period of time. State the assumptions for which this model is valid, and show that the inter-arrival times for calls are given by a negative exponential distribution. If an average of 120 calls per hour is received, calculate the probabilities that in any one minute the number of calls is (a) exactly one, and (b) more than two. What proportion of calls will be separated in time by more than one minute?

Inevitably, subscribers have to wait before their calls can be passed on by the exchange. Assuming a server queue model, with Poisson input and negative exponential holding time distributions, determine the exchange service rate if the expected number of customers having to wait is not to exceed three. At this service rate what is the probability of having to wait? Determine the new service rate if this probability is not to exceed 0.25.

5. Derive an expression for the probability of there being n items in a system having c service channels $(0 < n < c)$ with negative exponential service time and Poisson arrivals. Show that the probability of the system being idle is given by

$$p_0 = \{(\Sigma_{n=0,c-1}\, \varrho^n/n!) + (\varrho^c/((c-1!)(c-\varrho n)))\}^{-1}$$

where ϱ = arrival rate/service rate.

A port to unload container vessels is being planned, having seven berths which will be continuously available. The arrival of ships is expected to be random with an average of six per day. The time each ship will occupy a berth is expected to be constant since ships are all of the same type. What berth time should be planned if 70 percent of ships are not to wait for unloading because all berths are occupied? If the average number of arrivals increased by one ship per day, what is the probability that a ship must wait for a berth?

6. A telephone exchange has calls for trunk lines at a rate of one per minute with a negative exponential distribution of intervals between such calls. The holding rate is similarly distributed with an average of three minutes. Calls for trunk lines cannot be held if all the lines are engaged and are therefore lost. Calculate the smallest number of trunk lines required at the exchange if no more than one in a hundred calls on average are lost. Assume now that a particular type of call is treated by the exchange equipment in five distinct phases by a single service channel. Each of these phases takes an average of 5 s to process a call but the actual time varies widely. If the average number of such calls is 45 per hour, determine the average time spent in processing a call for each of three types of service time distribution:

1. Erlang service.
2. Total service time exponentially distributed.
3. Constant service time.

5.8 FURTHER READING

Bunday, B.D. (1986) *Basic Queueing Theory*, Edward Arnold.

Cooper, R.B. (1981) *Introduction to Queueing Theory* (2nd ed.), Edward Arnold.

Kendall, D.G. (1951) 'Some problems in the theory of queues', *Journal of the Royal Statistical Society*, **B13**, 151–85.

Mars, P. and Narendra, K.S. (1987) 'Routing, flow control and learning algorithms', *IEE First International Conference on Telecom. Networks*, 78–83.

Mars, P. and Popplebaum, W.J. (1981) *Stochastic and Deterministic Averaging Processors*, IEE Research Monograph.

Mars, P., Narendra, K.S. and Chrystall, M.S. (1983) 'Learning automata control of communication networks', in *Proc. 3rd Yale Workshop on Adaptive Systems Theory*, 114–19.

Murdoch, J. (1978) *Queueing Theory: Worked Examples and Problems*, Macmillan.

Newell, G.F. (1982) *Applications of Queueing Theory* (2nd ed.), Chapman and Hall.

6 Linear programming

6.1 PRELIMINARY OBSERVATIONS ON OPTIMIZATION

These opening remarks must serve two quite distinct purposes. First they must introduce the notion of optimization by explaining the vocabulary in common use; by indicating the intention of what is meant by an optimization method or technique; and by describing what is available to those who wish to use optimization with definite reasons in mind. Secondly, they must open the discussion on one particular technique: linear programming. By combining these two descriptions into a single section helps to relate a general commentary to a specific method. This aids the reader in understanding the generalities of optimization by means of direct reference to a comparatively well-known method. It should also show the reader why linear programming is the way it is as a consequence of the basic nature of and philosophy behind optimization itself.

In simple terms optimization is a formal approach towards obtaining the best performance from a given situation. This presupposes that the situation is variable, alterable and controllable. In other words that it possesses a structure which can be described in logical coherent terms and that this description suggests ways in which the best or optimum conditions for operation can be replicated. Such situations are not difficult to imagine and are immensely varied in their composition. To illustrate one can think of the following:

1. The organization of a factory to achieve maximum productivity.
2. The flight plan of a manned spacecraft to minimize fuel consumption on a given mission.
3. The arrangement of security and safety procedures to minimize hazards and risks at a liquefied natural gas plant.
4. The design of a rocket engine to provide maximum thrust for given design parameters such as size, weight, rate of fuel consumption and designed lifespan.

In all the above one is dealing with given resources and a stated objective and one attempts to produce attitudes, procedures, hardware and software designs or some other formal plan that achieves the objective within the resources available. In achieving the optimum, the plan will involve some term or combination of terms to take on an extreme value that will be either a maximum or minimum, e.g. maximum productivity, minimum fuel consumption, minimum risk or maximum thrust. In each case the objectives are all very sensible and highly desirable. They all represent logical extensions of basic common sense applied to situations where there is intense motivation to extract the best performance from situations. This then is the kernel to optimization philosophy: to achieve, by formal means, a performance in the extreme that is recognizably desirable in view of irresistible pressures, and which is indentifiable with the sense of purpose for which the situation prevails. This rather elaborate statement is helpful since one can question whether optimization for a given situation is needed and if so what form it should take in regard to the statement's applicability to the situation in hand. For example, under certain circumstances, the 'pressures that will not go away' might, unexpectedly go away; or, 'the sense of purpose for which the situation prevails' may call for minor or indeed trivial action to achieve the extreme performance. The watchword in all cases is 'pragmatism' and the policy is 'to defer if possible'. When committed to optimization, in the formal sense, it is essential to move swiftly. If commitment, in terms of resources, is underestimated the consequences can be fatal.

Proceeding therefore from this basic view of the notion of optimization one is faced with one of two general situations, each of which involves a resource, typically money, e.g. profit, budget, cost, and a figure-of-merit or 'alternative resource', e.g. performance. In the first of these two situations the resource is specified, i.e. a given budget, and the problem is to optimize the alternative resource, e.g. optimize performance, maximize reliability or minimize risk. In the second of the two, the performance or alternative resource is specified and the problem is to minimize the amount of resource, e.g. capital required to meet the specification of performance.

Turning now to an introduction of linear programming, one immediately defines, for the new vocabulary, the terms 'constraints' and 'objective function'. The notion of optimization in this specific situation is to maximize or minimize the objective function and simultaneously satisfy all of the constraints. These new terms highlight the need for formalism in optimization, in the sense that a description of the system requires, for analysis purposes, a logical, in fact mathematical, statement of the situation to be made which will precisely define the system of constraints and the objective function. Mathematically therefore the system can be represented by a set of variables each standing for a physical quantity or decision and the system of constraints and the single objective function or equations in these variables. There are certain basic properties to linear programming which can only be taken advantage of if the situation to be optimized meets certain preconditions:

1. All variables must be non-negative.
2. Each and every constraint must be linear, i.e. involving purely linear terms of the system variables.
3. The objective function is a linear equation, the subject of which is a single variable standing for system performance or the alternative resource, e.g. profit, wastage, physical size.

Very often the physical situation that becomes the object of optimization can be described or modeled as a system suitable for treatment by the linear programming method. In such cases the meaningfulness and value of optimization is not at all in doubt. The benefit that this technique provides to the managers of the situation has been and continues to be inestimable. Linear programming has been applied successfully many times and several bibliographies may be consulted to verify this.

In a great many situations the preconditions appertaining to the application of linear programming are not present, i.e. there are non-linear constraints or objective function. Alternatively, they are present but so also are other conditions that make the simple form of linear programming inadequate, e.g. integrality of the system variables. Fortunately the set of mathematical programming tools is not restricted to linear programming. Indeed a great many tools suitable for the treatment of non-linear and integrality conditions are available, proficient and widely used. They are continually being improved, extended and supplemented. (For a survey of what is contemporarily available, consult the references.) It is undoubtedly true, however, that linear programming, still widely accepted as the paramount optimization tool in the commercial sense, gave the much-needed start to this field of study and, as such, it makes sense to begin with this before looking at more sophisticated arrangements.

6.2 AN ELEMENTARY PROBLEM IN THE CHOICE OF TWO VARIABLES

It is accepted practice for 'beginners' to encounter a 'typical' problem as a worked exercise at an early stage in learning a new technique. Linear programming (LP) itself is most easily presented by means of the simplest of problems imaginable, i.e. one involving the determination of two variables.

Two types, A and B, of integrated circuit package are to be mounted on a card. Power delivered to the card must not exceed 2 W while the total current drawn should not be greater than 120 mA. Weight is an important consideration and the maximum package weight for this card is 16 g. One package type (A), per package, dissipates 0.1 W, draws 12 mA, weighs 2 g and occupies 3 cm^2. The equivalent characteristics for B are 0.2 W, 10 mA, 1 g and 2 cm^2 respectively. Assuming total package area to be the same as card area, what is the ideal combination of packages to maximize the card area in view of these constraints?

Since we need to determine the number of packages of each type, these represent unknown variables whose values are to be determined. Let the numbers of type A be a, and b the number of type B. Using these variables it is possible to establish the constraints (or inequalities) for the three conditions: power, current drawn and weight. Similarly the objective function, in this case package area, can be formulated.

The power consumed by each of the type A packages is 0.1 W and therefore a of these will take $0.1a$ W. The power taken by b of type B will be $0.2b$ W. This combination will consume a total power of $0.1a + 0.2b$ and it is stated that this should not exceed 2 W. This leads to our first mathematical constraint:

$$0.1a + 0.2b \leqslant 2$$

from which will be observed its linear form. To understand this constraint in graphical form requires little effort. First both a and b must be non-negative which restricts combinations of a and b plotted against each other to the first quadrant.

The line:

$$0.1a + 0.2b = 2$$

denotes a boundary forming an upper limit on the combinations of a and b which satisfy the power constraint. Therefore the region which defines a feasible combination of a and b, from the point of view of the power constraint alone, is the shaded triangular area in Figure 6.1, including the boundary itself. By identical reasoning, linear mathematical constraints can be formulated from the considerations for current and weight requirements giving:

$$12a + 10b \leqslant 120 \text{ and } 2a + b \leqslant 16$$

If the boundary lines for these two regions are plotted, the shaded portion now denotes the feasible region for combinations for a and b which satisfy all the constraints. Simultaneously, our search for values of a and b is confined entirely to this region including its boundary. The linear mathematical constraints define the region of feasibility wherein lies the optimal solution. To determine where the latter actually lies requires us to formulate the objective function which in this case is package (or card) area, denoted by Z and given by

$$Z = 3a + 2b$$

There are two ways in which this expression may be regarded mathematically:

1. It is a function of the two independent variables a and b. Thus $Z = Z(a,b)$.
2. Z is a parameter, specific values of which determine straight lines with

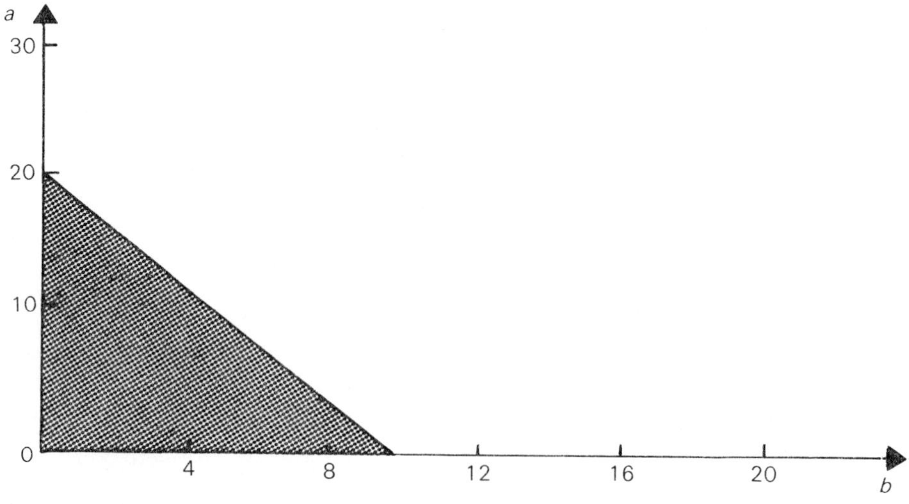

Figure 6.1

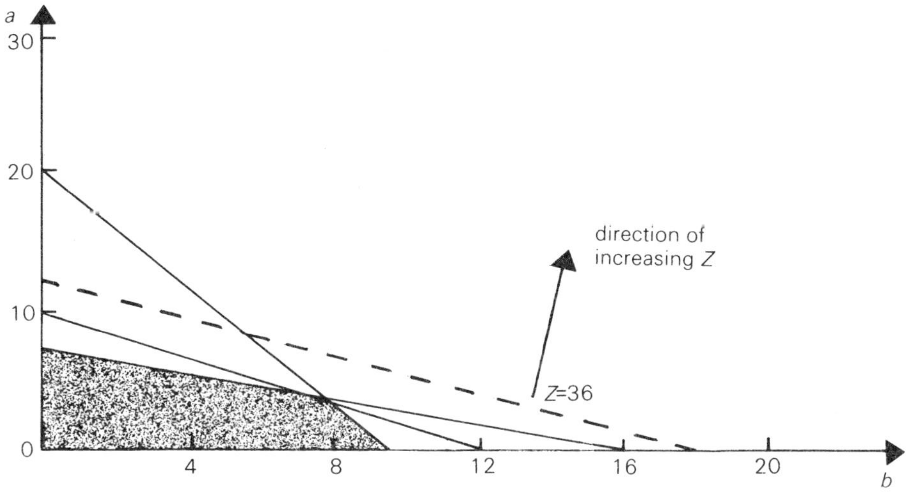

Figure 6.2

constant slope forming those values of a and b which satisfy the objective function accordingly.

To explore the region of feasibility it is preferable, by this graphical method, to refer to the objective function by means of 2 above. Thus the line for $Z = 36$ is shown in Figure 6.2 in relation to the feasibility region as a dotted line and the direction of increasing Z is perpendicular to this line and away from the region of feasibility. It is clear that values of $Z > 30$ produce combinations of a and b which are not feasible. Though we wish to maximize Z we must seek a value less than 30. We do this by moving in a direction perpendicular to the Z-line slope and towards the feasibility region. It is unnecessary to move into the region since the boundary of the region meets the required conditions.

The optimal solution will be at the point where a Z line touches the region. Here, the point lies at the intersection of the current and weight lines at which $a = 5$ and $b = 6$. Thus the maximum package area which can be obtained under the stated conditions is $27\,\text{cm}^2$.

6.3 WRITING LINEAR PROGRAMS

6.3.1 Introduction

Experience in writing computer programs is gained solely through writing programs. Self-evident as this seems it nevertheless is a constant reminder that by no other way than actual practice can one expect to gain proficiency and indeed fluency. The distinction between computer programming and mathematical programming has already been made. However what is true for computer programming as regards gaining experience is true also for linear programming, i.e. practice makes perfect. In essence, computer programming consists of the following:

1. Problem formulation, e.g. composing the flowcharts which define the problem and method of solution.
2. Knowledge of the computer programming language(s).
3. Skill at coding.

By comparison the writing of linear programs comprises the following:

1. Problem formulation – identifying the constraints and objective function.
2. Knowledge of available implementations on computers, e.g. computer packages.
3. Skill at coding, i.e. representing the constraints and objective function precisely, accurately and generally.

To assist the reader in gaining the necessary preliminary experience at writing linear programs, this section contains the following items: the linear program for exercise 6.1 (section 6.3.2); the linear program for exercise 6.3 (section 6.3.3); linear programming algebra (section 6.3.4); and a problem in network synthesis (section 6.3.5).

6.3.2 The linear program for exercise 6.1

At first sight this problem (from section 6.6, exercises) seems to bear little or no relation to that discussed in the previous section. In retrospect the reader should find that so far as linear programs are concerned there is virtually no difference.

The first point to note is that the problem posed is to find values for two unknown variables, namely tonnes of product A (a) and tonnes of product B (b). Secondly there are three immediately obvious conditions: mixing, firing and polishing. Finally the objective function is to maximize profit which is straightforwardly related to the values a and b.

Accordingly the mixing constraint is obtained by making the mathematical statement that the time taken to make a tonnes of product A (at 10 hours per tonne) and b tonnes of product B (at 5 hours per tonne) should not exceed 50 hours, i.e.

$$10a + 5b \leqslant 50$$

By a similar process we easily obtain the two further linear inequalities for the firing and polishing conditions:

$$6a + 6b \leqslant 36$$
$$4.5a + 18b \leqslant 81$$

The profit on each tonne of product A is £90 and on each tonne of product B is £70. The total profit denoted by Z is therefore given by:

$$Z = 90a + 70b$$

Noting further that both a and b must be non-negative the final linear program for this problem is maximize

$$Z = 90a + 70b$$

subject to: $10a + 5b \leqslant 50$, $6a + 6b \leqslant 36$, $4.5a + 18b \leqslant 81$ and $a \geqslant 0$, $b \geqslant 0$.

The solution of this linear program can be found in a manner identical to that used in the previous section, and is left to the reader as a formal exercise.

6.3.3 The linear program for exercise 6.3

This problem underlines the need to develop skills at problem formulation which is so necessary to the writing of linear programs if the technique is going to be successfully applied to the wide variety of situations where it is capable of optimizing systems performance. For this particular problem the choice of variables is crucial to the writing of the linear program. In fact that is only one choice to make, but how this is discovered is purely a matter of intelligent thinking, cultivated undoubtedly by the kind of experience which has been emphasized previously.

Since orders may be combined, combinations of width can be cut to length in such a way that resultant scrap is minimized and combinations of these various lengths satisfy the conditions on overall orders for the differing widths. These 'width combinations' must be all of those possible such that none leaves a scrap of 55 cm width or more. Table 6.1 shows the combinations possible.

If now we decide to make length cuts xi (in meters) of combinations C1 then the total length for each width must exceed or equal the limit established by orders for that particular width and the total scrap, area Z, must be minimized. Regarding first the 80 cm width, then:

$$x1 + 2x2 + x3 \geqslant 700$$

Table 6.1

	80 cm	70 cm	55 cm	Width scrap (cm)
C1	1	1	1	0
C2	2	0	0	45
C3	1	0	2	15
C4	0	0	3	40
C5	0	2	1	10
C6	0	1	2	25

For the 70 cm width we obtain:

$$x1 + 2x5 + x6 \geqslant 1{,}000$$

and for the 55 cm width we derive:

$$x1 + 2x3 + 3x4 + x5 + 2x6 \geqslant 700$$

These then are our three linear inequalities. The question now is 'what is the correct expression for Z, the area of scrap?' The first point to decide is what units of area to adopt since our lengths at present are in meters, whereas widths are in centimeters. If we choose m then the scrap from combination C2 is $0.45x2$. From combinations C3 to C6 inclusive the scrap areas are $0.15x3$, $0.4x4$, $0.1x5$ and $0.25x6$ respectively. There is no scrap from combination C1. The total scrap would appear to be therefore the sum of these five terms. However the solution to the linear program may provide lengths for the xi such that the combinations for the particular widths exceed the actual orders. If so there will, in such cases, be a contributary scrap area term of:

$$\text{excess length} \times 0.01 \times \text{cm width}$$

These excesses, if any, must be determined. Looking at our first constraint we observe that the value of the LHS is stated, formally, as being greater than (in excess of), or equal to, the RHS. Wherefore:

$$x1 + 2x2 + x3 - x7 = 700$$

where $x7 \geqslant 0$. In this case $x7$ represents the excess for the 80 cm width which will, of course, be zero if $x7 = 0$.

We shall see later that $x7$ will be referred to as a *slack variable* but for the moment we may easily observe that $x8$ and $x9$ may be regarded as terms denoting any excesses, if any, for the 70 and 55 cm width orders. Therefore to be accurate in formulating the expression for the scrap area, we must include the three terms $0.8x7$, $0.7x8$ and $0.55x9$. Wherefore:

$$Z = 0.45x2 + 0.15x3 + 0.4x4 + 0.1x5 + 0.25x6 + 0.8x7 + 0.7x8 + 0.55x9$$

The full linear program therefore is minimize:

$$Z = 0.45x2 + 0.15x3 + 0.4x4 + 0.1x5 + 0.25x6 + 0.8x7 + 0.7x8 + 0.55x9$$

subject to:

$$x1 + 2x2 + x3 - x7 = 700$$
$$x1 + 2x5 + x6 - x8 = 1{,}000$$
$$x1 + 2x3 + 3x4 + x5 + 2x6 - x9 = 700$$

and $xi > 0$ for $i = 1, 2, \ldots, 9$.

A solution by an algebraic method is left as an exercise. Section 6.4 describes the simplex method, this being the most prominent among those to have been implemented to provide computer solutions to linear programs.

6.3.4 Linear programming algebra

This section is primarily intended to consolidate the reader's familiarity with the algebra that so obviously relates to the formulation of linear programs.

The general linear programming problem is to maximize z for non-negative, $x1, x2, \ldots, xn$ subject to:

$$z + a01x1 + a02x2 + \ldots + a0nxn = b0$$
$$a11x1 + a12x2 + \ldots + a1nxn = b1$$
$$\ldots$$
$$am1x1 + am2x2 + \ldots + amnxn = bm \qquad (6.1)$$

It also includes the case of inequality constraints since these can be made into equality constraints by the introduction of new non-negative variables which do not enter z, the objective function. For example:

$$a01x1 + a02x2 \geqslant b0$$

becomes

$$a01x1 + a02x2 - x3 = b0$$

where

$$x3 = -b0 + (a01x1 + a02x2)$$

and

$$x3 \geqslant 0$$

Likewise,

$$a11x1 + a12x2 \leqslant b1$$

becomes

$$a11x1 + a12x2 + x4 = b1$$

where

$$x4 = b1 - (a11x1 + a12x2)$$

and

$$x4 \geqslant 0$$

The variables $x3$ and $x4$ are known as *slack* variables. It is assumed that $m + 1$, the number of equations, including the objective equation, is not greater than $n + 1$, the number of variables, including z as a variable. It is then possible to solve equation (6.1) for any $m + 1$ of the variables in terms of the $n - m$ other variables. Consider:

$$\begin{aligned} &z + 10x1 + 3x2 + 2x3 = 5 \\ &6x1 + 3x2 + 4x3 = 8 \\ &2x1 + 2x2 + x3 = 4 \end{aligned} \tag{6.2}$$

i.e. four variables and three equations.

This set can be rewritten as:

$$\begin{aligned} &z = -17/3 + (16/3)\cdot x3 \\ &x1 = 2/3 - (5/6)\cdot x3 \\ &x2 = 4/3 + (1/3)\cdot x3 \end{aligned} \tag{6.3}$$

The problem is exactly the same but presented in a slightly different form. A solution to this is now possible in terms of $x3$.

In general, if a set of $n-m$ variables can be given specified values then the other variables $z, x1, x2, \ldots, xm$ will be determined. If these values are chosen so that the $xi > 0$ for all i, then the solution is admissible or feasible, but they must also be chosen so that z is a maximum. The set (6.1) can be transformed into:

$$\begin{aligned} &z = b0' - (a0, m + 1\ xm + 1 + \ldots + a'0, nxn) \\ &\ldots \\ &xm = bm' - (am', m + 1 + \ldots + a'm, nxn) \end{aligned} \tag{6.2}$$

Knowing the coefficients $a'0k$, $k = m + 1, n$ to be positive, if any of the

xk, which are non-negative, are also non-zero, the effect is to reduce z. Further, knowing the bj', $j = 0, m$ to be positive, the xi, $i = 1, m$ take positive values when the non-negative xk are all zero. The set (6.4) therefore represents the solution to (6.1) and the problem as originally presented becomes that of obtaining the transformation of (6.1) to (6.4).

The reader should confirm that the general form of the linear programs for the integrated circuit package problem and the product mix problem are, respectively:

$$\begin{aligned} &\text{maximize } x \text{ where } z - 3a - 2b = 0 \\ &\text{subject to} \quad 0.1a + 0.2b + w = 2 \\ &\qquad 12a + 10b + x = 120 \\ &\qquad 2a + b + y = 16 \\ &\text{with } a, b, w, x, y > 0 \end{aligned} \tag{6.5}$$

and

$$\begin{aligned} &\text{minimize } z \text{ where } z - 90a - 70b = 0 \\ &\text{subject to} \quad 10a + 5b + r = 50 \\ &\qquad 6a + 6b + s = 36 \\ &\qquad 4.5a + 18b + 6 = 81 \end{aligned} \tag{6.6}$$

6.3.5 A practical problem in electrical network planning

There are many practical problems to which linear programming has been applied with great profit. The reader is referred to the several interesting texts and bibliographies which detail the range of this diversification which has contributed substantially to the appeal and success of this particular tool. A problem of particular interest to the author is that of logical design of electrical networks, where the optimal synthesis of an electricity supply network can be achieved to a large degree by the linear programming approach. A principal work of reference in this field is that by U.G.W. Knight (1958), and for precise details of this work the reader should consult the reference. Subsequent to this work a great deal of interest in this specific topic has arisen and several references are made to it in various sections of this text. This section, however, is concerned with the very basic definition of the problem in terms of deriving a linear programming solution.

A power supply network is designed so that it will transmit given amounts of electrical power subject to a number of conditions, which include the following:

1. The cost of the network to construct and operate should be kept to a minimum.

2. The continuity of supply afforded by the network should not be less than the minimum acceptable for the size and type connected to the network.
3. In the case of a generating station the connections provided should give adequate capacity out of the station under those circuit outage and load conditions assessed as technically and economically justified during the design study.
4. The technical design of the network should be adequate.

The final condition is one which is most sensible to apply to a suggested or outline design. As for the first three, these are amenable to a linear programming formulation. Consider the elementary situation of a single generating station and three load substations shown in Figure 6.3.

The variables for this problem relate to the number of circuits to be placed on the wayleaves (between pairs of stations) of which there are six, i.e. (1, 2), (1, 3), (1, 4) {direct connections from generation to load}, (2, 3), (2, 4) and (3, 4). Let these be denoted by x, $x2$, $x3$, $x4$, $x5$ and $x6$ respectively. Clearly $xi > 0$ but also it is necessary to limit the number of circuits on any wayleave which provides the additional constraints $xi < w$. When writing the security of supply conditions it is not possible to have an a priori knowledge of what the final network connections will be. It is therefore to specify minimum connections to every possible group of substations in order to ensure that all groups in the final design will have adequate circuit capacity connected to them.

Assuming that substations on the network to be designed are approximately equally loaded, the security of supply conditions can be specified as follows:

1. Each load substation must have at least $h1$ circuits connected into it.

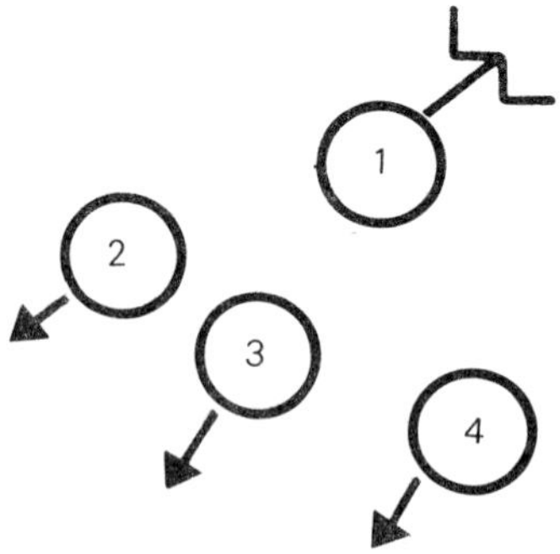

Figure 6.3

2. Each possible group of two load substations must have at least $h2$ circuits connected into it.
3. Each possible group of three load substations must have at least $h3$ circuits connected into it; and so on for every possible group of load substations of all sizes up to and including all the load substations. This produces then for our simple network design:

$$x1 + x4 + x5 > h1$$
$$x2 + x4 + x6 > h1$$
$$x3 + x5 + x6 > h1$$
$$x1 + x2 + x5 + x6 > h2$$
$$x1 + x3 + x4 + x6 > h2$$
$$x4 + x5 + x2 + x3 > h2$$
$$x1 + x2 + x3 > h3$$

Another important engineering consideration is to limit the number of circuits controlled at load substations, since this helps to reduce substation costs as a direct result of reduced capital investment in switchgear. For our small problem this gives rise to:

$$x + x4 + x5 < k$$
$$x2 + x4 + x6 < k$$
$$x3 + x5 + x6 < k$$

where k is a parameter whose value is specifically selected from a consideration of the type of proposed network, as also for $h1$, $h2$ and $h3$.

Finally the objective function can be formulated from a number of important criteria, chief of which is possible cost. If it can be assumed that wayleave costs are linear and would include a proportion of substation costs, e.g. controlling switchgear, establishment and civil engineering costs, then a suitable objective function for our simple network is:

$$z - c1x1 + c2x2 + c3x3 + c4x4 + c5x5 + c6x6$$

where the cs are the relevant per unit circuit costs for the respective wayleaves.

6.4 GENERAL SOLUTION OF LINEAR PROGRAMS

The simplex method provides a relatively efficient way of finding the m variables which if set (6.1) are solved in terms of them, gives the appropriate set (6.4). This set is known as a canonical form with respect to the basis $(z, x1, x2, \ldots, xm)$. The variables in the basis are known as basic variables. A basic solution to the equations is found by putting $n - m$ variables equal to zero. If z is in the basis and a basic solution is found in which the non-zero xs are non-negative, the solution is said to be feasible.

Let us look at the operation of this in relation to the general form of the linear program for the integrated circuit package problem, set (6.5), to give:

$$a = 8 - b/2 - y/2$$

Substituting for a in the remainder of the set (6.5) gives the following set:

$$\begin{aligned} z &= 24 - (-0.5b + 1.5y) \\ w &= 1.2 - 0.15b + 0.05y \\ x &= 24 - 4b + 6y \\ a &= 8 - b/2 - y/2 \end{aligned} \tag{6.7}$$

Now $z = 24$, $w = 1.2$, $x = 24$ and $a = 8$ is the new basic feasible solution, and the new basis is (z, w, x, a). This again is not optimal because of the negative coefficient of non-basic variable b. This variable may enter the basis but if it did so with $y = 0$ it could not exceed 8 or else w would become negative (second equation of set (6.7)). Also b could not exceed 6, with $y = 0$, or else x would become negative (third equation of set (6.7)). Therefore if x were to become a non-basic variable, then b would enter the basis. Rewriting the third equation of set (6.7):

$$b = 6 + 1.5y - x/4$$

and substituting into set (6.7) gives:

$$\begin{aligned} z &= 27 - (0.125x - 0.7y) \\ w &= 0.3 - 0.047x - 0.275y \\ a &= 5 + 0.125x + 1.25y \\ b &= 6 - 0.25x + 1.5y \end{aligned} \tag{6.8}$$

By inspection of the first equation of this set we note that this new basic feasible solution is also optimal. By setting $x = y = 0$ the optimal solution is found: $z = 27$, $a = 5$, $b = 6$. Notice that the non-basic variables x and y are also slack variables. Since these both take zero values this indicates zero slack for these conditions, i.e. current-drain and weight. The value of $w = 0.3$ indicates a slack of 0.3 W in the power constraint.

From the foregoing, the following general pattern emerges:

1. Find a basic feasible solution.
2. Check the coefficients in the expression for z. If they are all positive the solution is optimal.

3. If not, move to a new basis by taking one variable out of the basis and replacing it with a non-basic variable, as follows:
   (a) *choice 1:* basic variable to leave – that which ensures the new basic solution is feasible;
   (b) *choice 2:* non-basic variable to enter – that which gives an increased value of z at the new basic solution.

The following is a formal algebraic statement of the basic simplex procedure for the general linear programming problem.

1. Select a negative $a'0, j$ $(j = m + 1, \ldots, n)$. It is good practice, but not necessary, to select the most negative. The xj enters the basis.
2. Calculate $b'1/a1j$, $b'2/a2j$, $\ldots$, $b'm/amj$ for all i for which $a'ij \geqslant 0$ $(1 = 1, 2, \ldots, m)$. If none of the $aij \geqslant 0$ it is possible to increase xj indefinitely and the solution is unbounded. Otherwise, find the smallest, say $b'i/a'ij$, which then points to the basic variable in the ith equation, i.e. the $(i + 1)$ the basic variable, z being the first.
3. Solve the equations with new basis

$$z, x1, x2, \ldots, xi - 1, xj, x1 + 1, \ldots, xm$$

This involves dividing the ith equation by aij, to give:

$$\begin{aligned} xj &= b'i/a'ij - (a'i, m + 1/aij)\cdot xm + 1 \\ &= \ldots + xi/a'ij + \ldots + (a'in/a'ij)\cdot xn \end{aligned}$$

The existing jth basic variable must then be eliminated. This elimination will transform the system of equations and provide a new basic feasible solution. This can be checked for optimality. The process returns to step 1 if an optimal solution is still not found.

6.5 LINEAR PROGRAMMING AND THE THEORY OF GAMES

We conclude this chapter by noting how the linear programming technique can be used to solve a certain class of problem which John von Neumann posed in his original formulation of the theory of games. Whilst at first sight this may conjure up the games which people play for amusement, relaxation or exercise, what von Neumann had in mind was the more serious side, such as contests between nations, in which skilled adversaries pitted their resources against each other in order to win some advantage. The formulation of this theory is far from complete but the relationship of this theoretical scenario to the practical problems with which the systems engineer is confronted must be self-evident.

The practical problems to which games theory is applicable are those in which there is a conflict of interests and the participants have some control over the outcome. Anticipating the fields of interest of a systems team, the opposing interests may be those of competitors in business, opponents in war, competing subsystems for energy, space, and/or maintenance routine, or the managers and works in a system.

6.5.1 Defining the game model

The first step in defining the game model is to analyze a conflict situation with a view to identifying the parameters which can be symbolized, manipulated and developed to give the model its framework. The simplest conflict situation is a game itself and the first likely parameter we encounter is the number of players. In a game of bridge there are four players but two sets of pairs or partners, so the competition is really between two sides. Commonly, there are five players in a game of stud poker but it has been known for alliances to form and fall during the course of a game, thus disguising the real number of competitors. It is important at the outset to distinguish between persons as players and 'persons' representing a mutually conflicting interest with other 'persons'.

One of the fundamental distinctions in games' theory is the number of competing interests. The form of analysis and the entire character of the situation depend on this number. We concentrate initially on two-person games. A second important parameter to establish in the character of a game is the payoff, i.e. what happens at the end of the game? In most games there is usually just an exchange of assets, and very often the sum of gains, i.e. what persons win, is equal to the sum of losses. Mathematically this is represented by: $\Sigma g_i = \Sigma Z_j$ where G is the set of persons who gain, L is the set of persons who lose, g_i is the gain of person i in the set G, Z_j is the loss of person j in the set L. For completeness: $G \cap L = \phi$ and $G \cup L = E$, the set of all players. Although in many instances the game itself takes a percentage, i.e. for hospitality or home-profit, this losses-equals-gains rule is adopted throughout our treatment of games theory and so we are therefore concerned only with zero-sum two-person games.

Now that we have got our two players and the rule that the one's losses are the other's gains, our next step is to consider how the two play the game. This is concerned not so much with the roles or mechanics of the actual game but more with the plan that a player adopts so as to be more successful in terms of pay-off than his (or her) competitor. Strategy is defined as a plan so complete that it cannot be upset by enemy action or nature. At first sight this looks a panacea but the definition is a technical one relating precisely to the game model. Each player must have any number of strategies, but for convenience we only allow each player a finite number. The size of the game is governed by the number of strategies each player has. Let the players be X and Y (for you), then if you have three strategies and X has five, the game is 3 by 5 (or 3 × 5). What is of interest at this point is what happens when two players make a choice of strategy. This can be represented by a game matrix as shown in Table 6.2.

The intersection of any choice of strategies, e.g. Y3 and X4, refers to an element of the matrix. When players or strategies compete the game results in a gain for one player and the appropriate loss to the other. The amount (of exchange) could be made equal to the corresponding game matrix element with a sign convention that gives the direction of exchange, e.g. positive denotes you gain, negative denotes you lose and therefore X gains. For example: Y1 versus X5 − X gains 3; Y3 versus X4 − Y gains 8.

The play for the game is therefore entirely governed by the game matrix which symbolizes the result for all possible combinations of strategies. All the relevant information to the situation for which the game matrix is the model is contained therein so in effect the game may commence. We must reflect, however, on the effort that is needed to furnish a game matrix for the situation concerned. This is illustrated in simple fashion in later sections. In an actual systems engineering study it may typically be 90 percent of the entire effort in the competitive aspects of design.

If the game matrix governs the play of the game, it is common sense which satisfies the competitors' response to it. The sensible object of the player is to gain as much as possible from the game as he can, safely, in the face of a skillful opponent who is pursuing an antithetical goal. Game theory is therefore conservative in its make-up; it credits an opponent with intelligence. This being so, consider a two-person zero-sum game in

Table 6.2

		X				
		1	2	3	4	5
Y	1					−3
	2					
	3				8	

which all the payoffs (game matrix elements) are positive. The game is clearly unfair to X since Y always wins, but the arguments used for this arrangement will nevertheless appertain to the general case.

Interpreting Y's object, he wishes to act in such a manner that the least number he can win is as great as possible, irrespective of what X does. X should therefore act such that the most he can lose is as little as possible irrespective of Y's action. For either to depart from this will lead to one or the other obtaining less or losing more, as the case may be, than they actually needed to. In order to illustrate this, consider the following game matrix which is a model for a conflict situation involving two armies (or their commanders in this case). X prefers to fight the battle at a low altitude because his army has heavy fighting equipment which would be difficult to carry to great heights. Y prefers the high ground because his men do not have the same equipment but would be helped by the terrain at the higher levels. The game matrix elements indicate altitudes at route intersections in the battle areas. The situation resolves itself by X making a choice of the N–S routes that cross the particular terrain, whereas Y has the choice of E–W routes. The illustrations may appear slightly far-fetched but nevertheless serve the purposes of a game play.

Both X and Y know of the other's strategies and the game matrix, but not what the other will choose (Table 6.3). In Y's case the lowest altitudes for each of his strategies are 1, 2, 3 and 1. For X's strategies the highest altitudes are 7, 3, 5 and 6. In Y's case the worst (or lowest) altitude he can expect is 3 when he plays X2. In fact when Y3 is played against X2 each is doing the worst he can and can only do better if the other changes, e.g. if X changes, Y can expect 4 or 5. If Y changes, X can expect 2. Y does not risk playing Y1 or Y4 when he might get 7 or 6 since he might also get 1, namely, Y1 v. X4, as opposed to Y1 v. X1, and Y4 v. X4 as opposed to Y4 v. X3. Given that each will play safely crediting the other with intelligence, the result in this case is a foregone conclusion. To counter the query 'what about bluffing, double-bluffing, etc.?', consider the following:

Table 6.3

		X			
		1	2	3	4
Y	1	7	2	5	1
	2	2	2	3	4
	3	5	3	4	4
	4	3	2	1	6

Y: I am definitely going to play 3, i.e. safe.

X: Y is going to play safe. I will risk playing X4. I don't really mind losing 1 extra and Y may go for Y1 in the hope of getting 7 when I will then get 1.

Y: If X thinks of a risk in going for 4 then I won't want to go for Y1 because I'll end up at 1 so I'll go for Y4 and get 6.

X: Y may not go for Y1 if he thinks I'll risk X4, he'll then go for Y4. If so I'll end up at 6 – I don't like that. If Y does go for Y4 I'll go for X3 and still get 1 – I like that.

Y: X will know there is a chance of me getting 6 – he won't like that. He'll go for X3 and I'll end up at 1 again. If he goes for X3 I'll go back to 1, at least I'll get 5. Yes, that'll do.

X: Oh dear, Y may twig what I'm doing then he'll go for Y1. That's all right, I'll go for X4 and I'm still at 1.

Y: I seem to have been here before! (Moral: when in doubt, play safe.)

6.5.2 Non-Saddle-point Solutions

Consider the game shown in Table 6.4. If this game has a saddle-point then: max{min [a11, a12], min [a21, a22]} = min{max [a11, a21], max [a12, a22]}. The reader is encouraged to identify the saddle-points, if they exist, in the two games below:

3	6	6	5
5	4	5	4
(i)		(ii)	

The first of these games does not possess a saddle-point, in which case neither player may satisfactorily play a single strategy. For the game to be played it is obvious that each player will need to play both of their strategies in some combination. The question is – what combination?

Before moving to this we need some definitions. A pure strategy is one which is played all the time, e.g. in the case of a saddle-point game. A grand strategy is a mix of simple strategies. How then should we mix simple strategies?

Table 6.4

		X	
		1	2
Y	1	a11	a12
	2	a21	a22

Clearly it is important that X does not know what Y will do before Y plays and vice versa. Y can arrange for this by choosing a simple strategy at random. But does this mean that Y does not know himself what he is going to do? Or is Y able to govern the likelihood of the selection of a simple strategy whilst still keeping individual selections random? Suppose the probability that Y plays Y1 is p, then Y2 has probability $1 - p$. Similarly let X1 have probability q and therefore X2 is $1 - q$.

Suppose X plays strategy X1, then Y expects (using game (i)), $3p + 5(1 - p)$ per play of the game. If X2 is used, then $6p + 4(1 - p)$ per play of the game is Y's expectations. If Y is to play in a way that is unaffected by what X does, then these two expectations should be made equal, hence solving for p. Therefore:

$$3p + 5(1 - p) = 6p + 4(1 - p)$$
$$1 - p = 3p$$
$$p = 1/4;\ 1 - p = 3/4$$

Y should therefore play strategies Y1 and Y2 in the ratio 1 : 3 when on average he can expect to gain 4.5 per play. This notice is better than the maximum for Y. What should X do? He must act conversely, that is choose q so as to give the same expectations for Y1 or Y2, i.e.

$$3q + 6(1 - q) = 5q + 4(1 - q)$$
$$2(1 - q) = 2q \qquad \text{or } q = 1/2$$

X should therefore toss a coin before he plays! If he does he can expect to lose on average no more than 4.5 per play, i.e. what Y can expect to gain.

Two observations are appropriate here. The play of the game is unaffected by adding a constant to all payoffs or by multiplying all payoffs by a positive constant. The value of a game is affected when a constant is added to, or multiplied into, the payoffs. The reader should verify these by means of game (i), modified firstly by adding 5 to each payoff, then multiplying each payoff by 3. At this point it would seem reasonable to construct a game matrix by means of analysis of a conflict situation, albeit an artificial one.

Table 6.5

		X	
		1	2
Y	1	$a11$	$a12$
	2	$a21$	$a22$

A pair of bombers on a mission may come under attack from single-pass interceptors (i.e. they have only one chance of attack). One of the bombers carries the bomb while the other carries equipment to jamb radar and assess bomb damage, etc. The bombers fly such that bomber 2 is better protected than bomber 1. It is vital that the bomb is delivered, and the problem is in which bomber should it be placed? Additionally, which bomber should the interceptors attack?

This is clearly a 2×2 game as shown in Table 6.5. Y denotes bomber-strategy and X interceptor-strategy. The simple strategies are:

1. Y1 – bomb carried in bomber 1 (less favored position).
2. Y2 – bomb carried in bomber 2 (favored position).
3. X1 – interceptor attacks less favored position.
4. X2 – interceptor attacks favored position.

The payoffs will denote percentage survival of the bomb. If the game is played Y1 v. X2 or Y2 v. X1 then the payoffs must be 100 percent since the interceptors have chosen the wrong bomber to attack. If the game is played Y1 v. X1, the percentage survival is, say, 60 percent, whereas the Y2 v. X2 it is better, say 80 percent. Then the game would be as Table 6.6. It has no saddle-point. Adopting the convention above, Y should choose p (probability of employing Y1) be means of

$$60 + 100(1 - p) = 100p + 80(1 - p)$$
$$20(1 - p) = 40p$$
$$p = 1/3$$

The value of the game, i.e. what both Y and X can expect is 86.67 percent. (The ratio for X1 and X2 is also 1 : 2.) By employing a grand strategy over a pure strategy, Y increases the survival expectation from 80 to 86.67 percent, an improvement of about 8 percent. The playing of a grand strategy makes more sense with many missions but may be employed on a single mission if desired, although a pragmatic approach might be to adopt Y2.

Table 6.6

		X	
		1	2
Y	1	60	100
	2	100	80

Greater improvements than this are at stake depending on the payoffs and the reader is encouraged to calculate the improvement for game (iii) of using mixed over pure (Table 6.7). In attempting to enlarge on the 2×2 game, the immediate consideration is for games where one player has two strategies, whereas the other has more than two. These are referred to as $2 \times m$ or $m \times 2$ games.

6.5.3 Slightly larger games

Although $2 \times m$ games are interesting to deal with and there is a genuine desire to take on larger games, the first point that should be realized with these is that the initial move to make with them is to try to reduce their size! Consider the game in Table 6.8 where for the first time we see X with a chance to make a gain!

The first thing we look for is for strategies available to X that are clearly inferior to another. If so they can be discarded for X will never use them. For example, no matter what Y does, X is worse off to use 3 than, say, 2, where he gains for Y1 and Y2. The same can be said for 4, 5 and 7 when compared with 2. The same cannot be said for any other of X's strategies. The game therefore is reduced to that shown in Table 6.9. This is still not a 2×2 game and as it is 2×3, it must be solved as such. We must search for a solution using the 2×2 games that are implied by the 2×3 game which will satisfy the 2×2 game and which, incidentally, will then satisfy the original 2×3 game automatically. Y's two strategies compete against X's three and may do so two at a time (see Table 6.10).

The solution to this is $Y1 : Y2 = 9 : 5$

Table 6.7

		X	
		1	2
Y	1	5	1
	2	1	11

Table 6.8

		X						
		1	2	3	4	5	6	7
Y	1	−6	−1	1	3	7	4	3
	2	7	−2	6	3	−2	−5	7

Table 6.9

		X		
		1	2	6
Y	1	−6	−1	4
	2	7	−2	−5

Table 6.10

		X	
		1	2
Y	1	−6	−1
	2	7	−2

$$X1 : X2 = 1 : 13$$

and the value of the game is −19/14

What we must now try is Y's mix of 9 : 5 against X6. The value then is:

$$4*(9/14) - 5*(5/14) = (36 - 25)/14 = 11/14$$

Clearly X1 and X2 are a good mix of strategies for X; in fact they are the best and the reader should verify this. If therefore Y can do at least as well if not better against X's other strategies as he does against his best, then Y's mix against X's best strategies are those he should go for. The previous 2 × *m* game was solved as a 2 × 3 game finally, since it was possible to remove four of X's strategies. Only three were then active. The removed strategies were clearly inferior on all counts and it could be relied upon that they would never be called for. Without dwelling on why they were there in the first place, it is of great satisfaction to the 'gamesman' to see the size of the game reduced in this way. Smaller games are considerably easier to deal with.

The general principle that reduces the size of games in this way is termed dominance. This applies to either player having many strategies. A dominant strategy is one where the payoffs are favorable to Y for all of X's strategies compared with another strategy which is therefore the dominated one. If Y is the many-strategy player to which the dominance principle is being applied, the dominated strategies should be removed. They are inactive strategies. In the case of X being the many-strategy player, then the dominance principle states that the dominant strategies should be removed for they are X's inactive strategies. Applying dominance can certainly reduce the size of a 2 × m game, although one assumes of course that the two-strategy player has both strategies active. Nevertheless one still may be left with a large 2 × m game to solve. The rule is that the solution to the game is in effect the solution to one of its 2 × 2 games. The question is, which one? A graphical technique can be of some help here.

6.5.4 A graphical solution for 2 × m and m × 2 games

Take the last game and do not bother to apply dominance, for we shall discover that the graphical technique will eliminate automatically the inactive strategies of the many-strategy player. The technique is to plot the payoffs on separate vertical axes and connect pairs by straight lines according to the strategies of the many-strategy player (see Figure 6.4). Although this looks rather a complicated affair, a solution is near at hand. Remember that the vertical scale is payoff to Y, so Y must choose his mix of strategies, i.e. a point somewhere between the two axes, in such a way as to maximize his minimum gain. The minimum payoff to Y will be found on the envelope that forms the lower bound to the lines that

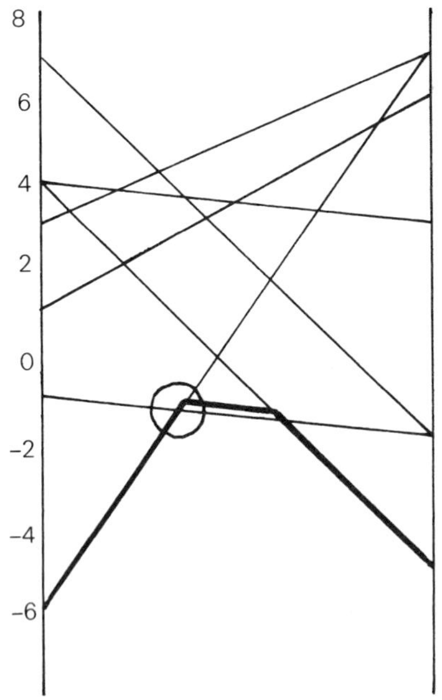

Figure 6.4

criss-cross Y1 and Y2. This is the heavy line. Clearly the maximum profit to Y on this envelope is at its highest point – marked with a circle.

This point is at the intersection of X1 and X2 which we already know from earlier calculations to be X's best mix. The envelope is formed by X1, X2 and X6 which are X's active strategies. All other strategies are irrelevant to the game and they turn out to be the dominant strategies.

The appropriate 2 × 2 game can now be solved, but even this is not really necessary. (What a change for the gamesman!) The point on the vertical scale is the value of the game since it is the payoff to Y (negative in this case and therefore a payoff in real terms to X). The point on the horizontal scale determines the ratio in which Y should mix his strategies.

The same technique may be used for an $m \times 2$ game where Y is the many-strategy player. Consider, for example, game (v) in Table 6.11. In this case the vertical axes in Figure 6.5 represent X's strategies X1 and X2 and the lines that criss-cross are Y's strategies. It should be noted therefore that vertical scale is being looked at by X, while the same sign convention is kept.

X now wishes to minimize the most he has to payoff to Y. The maxima of his payoffs to Y (though in effect gains to himself) are represented by an envelope that forms the upper bound to the criss-crossing lines, the strategies of Y. The lowest point on this envelope marks the payoff to X for which he settles and again is at the intersection of Y's best mix of strategies.

6.5.5 LP solutions of any finite zero-sum two-person game

Our discussion on games theory has largely been concerned with formulating a specific type of conflict situations, involving two competitors. There is little doubt that the systems team will need to do some forcing of their depicted problem in order to obtain the necessary approximations to a two-person game. Their task even then is not finished. The strategies

Table 6.11

		X	
		1	2
Y	1	4	−3
	2	2	1
	3	−1	3
	4	0	−1
	5	−3	0

(v)

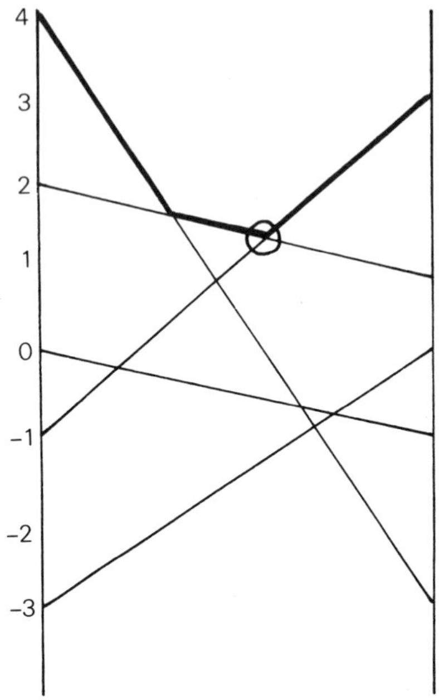

Figure 6.5

of the two persons need to be enumerated and the payoffs either calculated, estimated or guessed (in an educated manner, of course) for the different pairings of competing strategies.

At this point the game as formed may be solved. The first step is always to look for a saddle-point. If it exists the game is automatically solved. If not, the next step is to reduce the game size by dominance.

For games which remain large ($m \times n$) after all attempts to reduce them have been exhausted, solutions are generally obtained with recourse to computer packages. Indeed a game can be transformed into a linear programming problem, the computer solution of which is easily obtained, by suitable manipulation of the game matrix. Let this be represented by $A = aij$, then if player X chooses a mixed strategy ($x1, x2, \ldots, xn$) then he loses at most

$$\text{Max}_i \; \Sigma_{j=1,n} \; aij \cdot xj = v, \text{ say}$$

We have, therefore, for all i,

$$ai1\ x1 + ai2\ x2 + \ldots + ain\ xn \leqslant v$$
$$x1 + x2 + \ldots + xn = 1$$
$$xj \geqslant 0, j = 1, 2, \ldots, n$$

Similarly, if Y plays a mixed strategy ($y1, y2, \ldots, ym$) then he receives at least

$$\text{Min}_j \; \Sigma_{i=1,m} \; aij \cdot yi = w, \text{ say}$$

We have, therefore, for all j,

$$a1j\; y1 + a2j\; y2 + \ldots + 1mj\; ym \geqslant w$$
$$\Sigma_{i=1,m}\; yi = 1$$
$$yi \geqslant 0,\; i = 1, 2, \ldots, m$$

Now Y wishes to make w as large as possible. Written in this form, we have not yet reached a linear programming formulation since w need not be non-negative. However, we assume that we have added, to all elements of the game matrix, an amount large enough to make w positive. The optimal strategies thereby remain unaltered, while the value of the game is simply increased by the same amount and can obviously be eventually adjusted. If w is considered positive then we can divide the inequalities in the yi by w, and writing $\text{Y}i = yi/w$ we obtain

$$aij\; \text{Y}1 + a2j\; \text{Y}2 + \ldots + amj\; \text{Y}m \geqslant 1$$

for all j.

We now wish to maximize w, or conversely minimize:

$$\text{Y}1 + \text{Y}2 + \ldots + \text{Y}m\; (= 1/w)$$

subject to the above constraints and of course the non-negativity of the $\text{Y}i$. This is the linear programming problem in its usual form and of course, simultaneously, the dual problem is

$$\text{maximize } \text{X}1 + \text{X}2 + \ldots + (= 1/v)$$

subject to

$$ai1\; \text{X}1 + ai2\; \text{X}2 + \ldots + ain\; \text{X}n \leqslant 1$$

for all i and $XJ \geqslant 0, j = 1, 2, \ldots, n$. By solving one of these we implicitly solve the other also.

Having found the Xj and Yi and the minimum of $\Sigma_{i=1,m}\, Yi$ equals the

Table 6.12

	X		
	6	0	6
Y	8	−2	0
	4	6	5

maximum of $\Sigma_{j=1,n} Xj$, and is the value of the game v, the best strategies are indicated by Xjv and Yiv. As an example of how the linear programming method may be used to solve a larger game in this way, game (vi) is taken as an exercise. The game matrix in its original form is shown in Table 6.12.

To ensure a positive value of the game it is necessary only to add 2 to all the payoffs, giving:

8	2	8
10	0	2
6	8	7

Either of the two linear programming problems can be solved:

1. $8Y1 + 10Y2 + 6Y3 \geqslant 1$
   $2Y + 8Y3 \geqslant 1 \qquad Y1, Y2, Y3 \geqslant 0$
   $8Y1 + 2Y2 + 7Y3 \geqslant 1$
   Minimize $Y1 + Y2 + Y3$

2. $8X1 + 2X2 + 8X3 \leqslant 1$
   $10X1 + 2X3 \leqslant 1 \qquad X1, X2\ X3 \geqslant 0$
   $6X1 + 8X2 + yX3 \leqslant 1$
   Maximize $X1 + X2 + X3$

6.6 EXERCISES

1. A plant manufactures two products A and B. Three operations are involved in producing both A and B. These are designated:

(a) mixing M,
(b) firing F,
(c) polishing P.

The overall times available for these operations are 50 hours for M, 36 hours for F and 81 hours for P.

The times in Table 6.13, in hours per tonne, are required for the several operations on each product. Profit on each product per tonne is £90 for A and £70 for B. How many tonnes of each product must be manufactured to maximize the profit? Explain in detail your technique for computing this profit and state its value.

2. A manufacturer needs an alloy consisting of at least 40 percent tin, with remainder lead and zinc in any proportions. Five alloys are available and these are given in Table 6.14, together with their constituent metals and alloy cost. It is

Table 6.13

	M	F	P
A	10	6	4.5
B	5	6	18

required to blend these alloys to provide the least expensive product. Without attempting to solve, write the constraints and objective function for this problem.

Table 6.14

Alloy	Percentage lead/kg	Percentage zinc/kg	Percentage tin/kg	Cost £/kg
1	40	50	10	2.0
2	60	30	10	2.2
3	30	30	40	2.5
4	30	40	30	2.4
5	30	20	50	2.3

3. A float glass production line manufactures a glass ribbon 2054 cm wide. Customers orders on one day are for: 700 m for 55 cm width; 1,000 m of 70 cm width; 700 m of 80 cm width.

The orders are acceptable to customers in a number of lengths of the required width and so the manufacturer can combine the orders and cut them together. Attempt to solve the problem, using the simplex method, of minimizing the scrap which results from the cutting process. (Assume no breakages in the glass-cutting.)

4. A city is to defend against air attack by deploying a combination of low altitude and high-altitude guided missiles. The relative performance of these two types, depending on the nature of the enemy attack, is shown in Table 6.15. The maximum capability of the enemy for each kind of attack is also estimated.

Table 6.15

Type of missile	Altitude of attack			Cost per missile
	Low	Medium	High	(£10^3)
Low altitude	0.75	0.5	0.25	25
High altitude	0.25	0.5	0.25	50

Using the method of linear programming, determine the exact number of each type of missile which the city should deploy, in order to provide maximum capability against the enemy for the least possible cost.

5. Explain the minimax principle of game theory.

Two armies are fighting a war on mountainous terrain which is criss-crossed by passes. By virtue of their respective forces the red army chooses to fight on high ground while the blue army chooses the low ground. Battles are fought at the intersections of passes and the altitudes of these are indicated in Table 6.16. If the red army is to choose the E–W pass and the blue army has a choice of N–S routes, calculate the optimal strategies of the two armies. At what average altitude is the war fought?

Table 6.16 Altitudes at intersections

		Blue				
		1	2	3	4	5
Red	1	7	2	5	1	3
	2	6	1	4	1	2
	3	2	5	3	4	8
	4	1	4	2	4	6
	5	2	3	3	1	7
	6	5	2	1	4	3

6. A city must prepare its defences against possible attacks from an enemy country. The payoff matrix is shown in Table 6.17 indicating the various strategies for each opponent and the corresponding probabilities that the city will survive. Calculate the optimal strategies of the opposing sides and determine the average probability of the city's survival.

Table 6.17 Probability of survival (percentage)

		Enemy				
		1	2	3	4	5
City	1	70	20	50	10	30
	2	65	15	45	10	15
	3	20	50	30	40	80
	4	5	45	25	35	55
	5	20	40	15	25	50

7. A float glass production line is producing a large amount of breakages and the cross-cutters are suspected. The manufacturer (X) can either

(a) ignore the situation;
(b) perform some tests and make changes as necessary; or
(c) renew the cross-cutters.

A competitor, Y, has four possible outcomes open to him, to try to take advantage of manufacturer X's dilemma. A gain matrix is produced, Table 6.18, showing the gains to each in terms of their relative positions in the market. Use the minimax criterion to determine the optimal strategy of manufacturer X.

Table 6.18

		Competitor's outcomes			
		1	2	3	4
Manufacturer X	1	3	−4	2	5
	2	7	6	7	−5
	3	5	4	6	4

6.7 FURTHER READING

Beale, E.M.L. (1968) *Mathematical Programming in Practice*, Pitman.

Bunday, B.D. (1984) *Basic Linear Programming*, Edward Arnold.

Charnes, A., Cooper, W.W. and Henderdson, A. (1953) *An Introduction to Linear Programming*, John Wiley.

Gass, S. (1985) *Linear Programming, Methods and Applications*, McGraw-Hill.

Knight, U.G.W. (1958) 'The logical design of electricity networks using linear programming methods', *Proc. IEE*, **107**, 275–90.

Koopmans, T.C. (Ed.) (1951) *Activity Analysis of Production and Allocation*, John Wiley.

Luce, D.R. and Raiffa, H. (1957) *Games and Decisions: Introduction and Critical Survey*, John Wiley.

Manne, A.S. (1980) 'Linear programming and sequential decisions', *Management Science*, **6**.

Thie, P. (1979) *An Introduction to Linear Programming and Game Theory*, Wiley.

Vajda, S. (1961) *Mathematical Programming*, Addison-Wesley.

Vajda, S. (1981) *Linear Programming, Algorithms and Applications*, Chapman and Hall.

Walsh, G.R. (1985) *An Introduction to Linear Programming*, Wiley.

7 Dynamic programming

7.1 THE PRINCIPLE OF OPTIMALITY

Dynamic programming is a deceptively simple technique and therefore is often elusive to grasp. Presenting it for others to understand cogently and unambiguously poses a challenge in itself. Part of the problem may lie in the fact that it is essentially a very simple-minded approach to unlimitedly complex problems. It has been said that the principle upon which dynamic programming is based is a truism. Perhaps the reader would like at this stage to judge whether this is so. Consider a journey from A to D via B and C, Figure 7.1. We are concerned with two aspects of the journey, i.e. routes and points on the journey, and clearly these are inextricably related, e.g. the sub-routes AB, BC, CD, BCD, BED, ABCD and ABED are all clearly defined with respect to the points on each journey.

The suggestion is made that the sub-route ABCD is in some sense, e.g. scenically or economic on fuel, the *optimal route* from A to D (and presumably therefore no other such route is optimal, e.g. ABED cannot be optimal according to the same criterion). The fundamental principle upon which dynamic programming rests, the principle of optimality originated by Richard Bellman (1957), would suggest that if ABCD is optimal then regardless of the journey taken from A to B, having reached

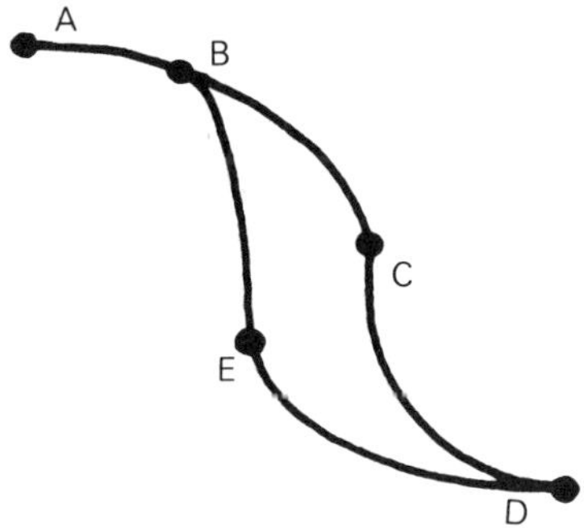

Figure 7.1

B it is essential that BCD is an optimal route in order for ABCD to be optimal. (This should be interpreted as a necessary but not a sufficient condition since clearly some comment is required on the optimality of the route between A and B.) The point that this principle makes is that in a series of decisions (effectively a sequence of points or succession of sub-routes), in order for an overall route to be optimal, the route leading onwards from any point reached along the optimal route must constitute an optimal sub-route with respect to such points. With care you will see that the same reasoning can be applied in the case of this illustration to routes 'from A' as opposed to 'to D'. This principle appears as a truism as evidenced by the simplistic manner in which it is proved by a converse argument. If BCD is not to be an optimal route from B to D (with respect to its start point B of course), then some other route, e.g. BED, must be optimal. If this were so then ABCD could not be the optimal route from A to D since ABED would be an improvement on this of necessity.

That argument should suffice for the time being and the reader is now advised to adopt the formal definiton of the principle of optimality which makes use of the term 'policy' interpreted as a sequence of decisions (a decision can be regarded as a choice of value of an unknown variable, or a next point in a sequence of such points, or a next sub-route, or whatever is most conveniently appreciated by the reader). The principle of optimality states that regardless of the initial state (point) or the initial decision (sub-route) of an optimal policy (route), it nevertheless has the property that all remaining decisions must constitute an optimal policy with respect to the state (point) resulting from the first decision.

The whole essence of the dynamic programming method is to reduce complex (multi-decision) optimization problems into a series of single-decision optimization problems where the optimization at each stage can be carried out either by calculus, for continuous functions, numerical techniques for discrete functions, or by some other appropriate means. Clearly, to be of advantage, the labor involved in solving n single-variable problems (effectively the same one each time) must be less than one huge attempt to solve an n-variable problem.

Let us look at a specific problem to judge whether dynamic programming will be of advantage. The problem is: 'How to divide 20 units of some resource among 4 competing parties in order to maximize the product of the allocations?' This can be thought of as a derivative of a simple reliability problem where four stages of a system, arranged in series, compete for resources in order to become more reliable. Overall system reliability is the product of stage reliabilities. Mathematically this can be written as:

(a) *Choose* values for x_i, $i = 1, 2, 3, 4$, so that $x_i > 0$ (any x_i) $= 0$ makes the product zero and therefore a minimum? *such that*
(b) $\Sigma_{i=1,4} x_i = 20$, *and*
(c) $z = \Pi_{i=1,4} x_i$ is a *maximum*.

The first question is a strange one to ask but one finds that it pre-empts in every single instance the formulation of any problem in dynamic programming terms. The question is 'what will the maximum value of z depend on?' Let the maximum value be denoted by f. Clearly f depends on the initial resource, in our case 20, but in general let us say this is X. It will also depend on the number of parties in competition for the resource, in our case 4, but in general say N. Strictly speaking therefore:

$$f = f_N(X)$$

Then, what is $f_4(20)$ and more to the point, what values of x_i will combine to yield a value for z of $f_4(20)$? We cannot determine $f_4(20)$ until we know something about $f_4(\mathrm{X})$ and moreover something about $f_3(X)$, $f_2(X)$ and $f_1(X)$. Let us start with $f_1(X)$. Clearly: $f_1(X) = X$. This is true for all X, so for example: $f_1(0) = 0$; $f_1(20) = 20$; $f_1(\mathrm{x}) = x$ and $f_1(X - x) = X - x$.

We can state unequivocally therefore that the solution to the problem of allocating to a single party is entirely known to us and on this basis we now build the solutions to larger-sized problems. It does not matter what our first decision is when two people are involved (or 2,000 for that matter), so let it be x, i.e. one of the two gets x where clearly $0 < x < 20$.

Now what we do with the rest must be done optimally (according to the principle of optimality). In fact we know how to do this, for with X the result is $f_1(X)$ so having given x to one, 20-x will be left and our result in this case will be $f_1(20 - x)$. If we had started with X shared between two, the best result we could get, depending on x, would be the product: $x \cdot f_1(X - x)$. It is a simple matter to choose x to make this a maximum, and when we do of course it will be equal to $f_2(X)$. Therefore:

$$f_2(X) = \max[x \cdot f_1(X - x)]; \; \{0 < x < X\}$$

since $f_1(X - x) = X - x$, the answer is easy to find using calculus and we get $x = X/2$, and $f_2(X) = X^2/4$. We can make use of this result to find $f_3(X)$. Remember this is the maximum value of z when X is shared among three. Again, no matter how much we give one of them, say x, the remainder $X - x$ must be shared among two to give the maximum value of z. We know how to do this, and the result is $f_2(X - x)$. Therefore:

$$f_3(X) = \max[x \cdot f_2(X - x)]; \qquad \{0 < x < X\}$$

This gives us $x = X/3$ and $f_3(X) = X^3/27$. It should not surprise the reader to know that

$$f_4(X) = \max[x \cdot f_3(X - x)]; \qquad \{0 < x < X\}$$

whence $x = X/4$ and $f_4(X) = X^4/256$.

Now if $X = 20$ then we give the first one 5 (whoever that is), leaving 15. Sharing 15 among three we give the first one 5, leaving 10. Notice that the first one of the threesome is identically the second of the foursome. This underlines the point that the optimization is a series of single (first) decisions of continually expanding problems. The fact that these decisions are imbedded in the largest (N-variable) problem is revealed as a consequence of the nature of the technique of converting the largest problem into N identical problems each of a single variable.

Returning to the allocation problem; sharing the remainder 10 between two we give each 5. It turns out that all four get an equal amount in order to solve the problem of maximizing z. But of greater importance is the fact that any size problem with any amount of initial resource can now be solved with ease. So what might have looked at first as unnecessary extra labor has repaid itself handsomely.

There are in fact many excellent examples of how dynamic programming may be applied to quite lifelike situations and which reinforce one's understanding of the method. The reader is encouraged to look at Boardman *et al.* (1986), Cooper and Cooper (1981), Norman (1975) and Bellman (1957) to consolidate that which has been gained here. However, a second example, is now given, and this particular one has a rather interesting corollary.

7.2 APPLICATION TO THE DISTRIBUTION OF GOODS

A transport manager must decide how to move 600 tonnes of material, having three modes of shipment open to him. Each of the modes, air, road and rail, make a charge proportional to the square of the load carried but their charges differ of course: the unit costs are 10, 5 and 10/3 respectively. The transport manager decides to move the material in such a way as to minimize the transportation costs. How is this to be done?

This is not dissimilar to our previous example and we have learned by now to generalize so that if we denote by f the minimum value of shipping costs, then we can write:

$$f = f_N(y)$$

where N is the number of modes of transport available to the manager and y is the load to be shipped. Let us say the unit cost of transport mode i is ai. Then f_1, whatever mode 1 is, will be given by:

$$f_1(y) = a_1 \cdot y^2$$

What is $f_2(y)$? Well no matter how much we allocate to mode 2, say x, then the remainder must be allocated at minimum cost. This will give a result of $f_1(y - x)$. The total cost will be:

$$a_2x^2 + f_1(y - x)$$

which depends on x. If this expression is minimized with respect to x, the result will give $f_2(y)$. Therefore:

$$f_2(y) = \min[a_2x^2 + f_1(y - x)]; \qquad \{0 < x < y\}$$

In full this is:

$$f_2(y) = \min[a_2x^2 + a_1(y - x)^2]; \qquad \{0 < x < y\}$$

which gives:

$$x = a_1y/(a_1 + a_2)$$

and

$$f_2(y) = a_1a_2y^2/(a_1 + a_2)$$

It is now important to realize that this approach can easily be extended to any number of modes of transport and any size of load, so the extra work in tidying up the algebra at this stage will be worth it in time to come. Let

$$f_1(y) = b_1y^2 \quad \text{so} \quad b_1 = a_1$$

and

$$f_2(y) = b_2 y^2 \quad \text{so} \quad b_2 = a_2 b_1/(a_2 + b_1)$$

We wonder now whether b_3 will be equal to $a_3 b_2/(a_3 + b_2)$, since that would fit the emerging pattern. If so, a general form can be obtained as was possible with the previous problem. We can only find this out by considering the formula for $f_3(y)$ which is:

$$f_3(y) = \min[a_3 x^2 + f_2(y - x)]; \qquad \{0 < x < y\}$$

or in full

$$f_3(y) = \min\,[a_3 x^2 + b_2(y - x)^2]; \qquad \{0 < x < y\}$$

Notice we have seen this problem somewhere before. In fact we have already solved it (twice!) but with different symbols. The result here is:

$$x = b_2 y/(a_3 + b_2)$$

and

$$f_3(y) = a_3 b_2 y^2/(a_3 + b_2)$$

which is rather nice, because we can now write down the general problem with its solution:

$$f_N(y) = \min[a_N x^2 + f_{N-1}(y - x)]; \qquad \{0 < x < y\}$$
$$N = 2, 3, 4 \ldots$$
$$f_1(y) = a_1 y^2$$

This general problem has for its solution:

$$f_N(y) = b_N y^2$$

and

$$b_N = a_N b_{N-1}/(a_N + b_{N-1})$$

and

$$x_N = (\text{for } N\text{th mode}) = (b_{N-1}/(a_N + b_{N-1}))\cdot y$$

It is always encouraging to find the answers to the original problem, having developed a general solution to the larger problem. Therefore by labeling the various modes of transport we effectively assign the a_1, hence the b_1 and hence determine the individual allocations. Thus if mode 1 is air, mode 2 is road and mode 3 is rail, then $a_1 = 10$, $a_2 = 5$ and $a_3 = 10/3$, therefore $b_1 = 10$, $b_2 = 10/3$ and $b_3 = 5/3$, so that $x(\text{rail}) = 300$, $x(\text{road}) = 200$ and this leaves 100 for the air. The minimum cost is therefore 600,000 units for the load of 600 tonnes.

If the reader has not already noticed the analogy, it may now be seen that the load is divided among the three modes of transport exactly in the same manner as a current divides in a parallel resistive circuit. Therefore just as the load (shipment in the original example) is divided so as to minimize the total cost (which is proportional to the square of the load), then a current divides in this type of circuit so as to minimize the total power loss. Full marks to nature!

7.3 A NETWORK PROBLEM AND NUMERICAL SOLUTION

The network shown in Figure 7.2 represents a series of towns across a particular county linked by major roads. The intention is to travel from town A to town N by the minimum-distance route where the distances traveled by major roads are indicated by the numbers appended to the links (or arrows) of the network model. It will become obvious that five major roads, five stages, are required to undertake the journey. The

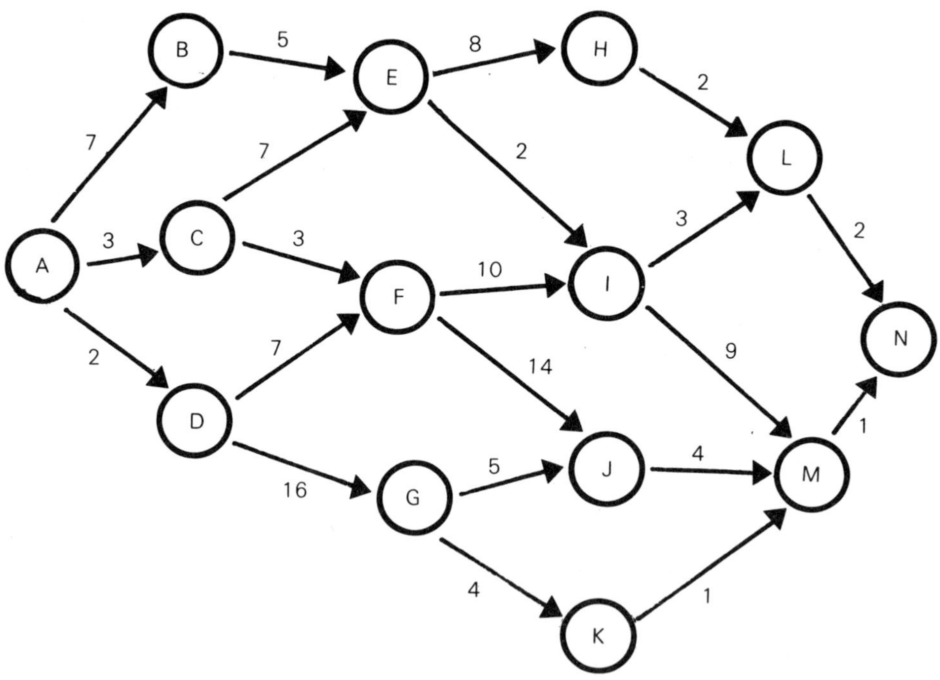

Figure 7.2

question is what decisions are to be taken at successive stages in order to travel by the minimum-distance route?

If N represents our destination, then L and M are stage 1 points (or towns), H, I, J and K are stage 2 points, E, F and G are stage 3 points, B, C and D are stage 4 points and A is the only stage 5 point.

Let us define $f_K(\alpha)$ as the minimum distance to N from point $\alpha(\alpha =$ A, B, C, ...) over K stages. We seek $f_5(\mathrm{A})$ and the decisions at each stage that produce it. We must solve single-stage problems first, then double-stage, and so on, up to $f_5(\mathrm{A})$.

We observe for each stage the $f_K(\alpha)$ and the route that gives us this value and we can indicate by numbers contained in squares, appended to the appropriate points (or towns), the corresponding values of $f_K(\alpha)$.

First stage:

$$f_1(\mathrm{L}) = 2:\mathrm{LN}$$
$$f_1(\mathrm{M}) = 1:\mathrm{MN}$$

Second stage:

$$f_2(\mathrm{H}) = 2 + f_1(\mathrm{L}) = 4:\mathrm{HLN}$$

N.B. The first decision of the two-stage problem is made for us, but the remaining decision, equally straightforward in this case, is made to yield an optimal value for distance henceforward.

$$f_2(\mathrm{L}) = \min[(3 + f_1(\mathrm{L})), (9 + f_1(\mathrm{M}))] = \min[3 + 2, 9 + 1] = 5:\mathrm{ILN}$$
$$f_2(\mathrm{J}) = 4 + f_1(\mathrm{M}) = 5:\mathrm{JMN}$$
$$f_2(\mathrm{K}) = 1 + f_1(\mathrm{M}) = 2:\mathrm{KMN}$$

Third stage:

$$f_3(\mathrm{E}) = \min[(8 + f_2(\mathrm{H})), (2 + f_2(\mathrm{L}))] = \min[8 + 4, 2 + 5] = 7:\mathrm{EILN}$$

$$f_3(\mathrm{F}) = \min[(10 + f_2(\mathrm{L})), (14 + f_2(\mathrm{J}))] = \min[10 + 5, 14 + 5]$$
$$= 15:\mathrm{FILN}$$

$$f_3(\mathrm{G}) = \min[(5 + f_2(\mathrm{J})), (4 + f_2(\mathrm{K}))] = \min[5 + 5, 4 + 2] = 6:\mathrm{GKMN}$$

Fourth stage:

$$f_4(\mathrm{B}) = 5 + f_3(\mathrm{E}) = 12:\mathrm{BEILN}$$

$$f_4(\mathrm{C}) = \min[(7 + f_3(\mathrm{E})), (3 + f_3(\mathrm{F}))] = \min[7 + 7, 3 + 15]$$
$$= 14:\mathrm{CEILN}$$

$$f_4(\mathrm{D}) = \min[(7 + f_3(\mathrm{F})), (16 + f_3(\mathrm{G}))] = \min[7 + 15, 16 + 6] = 22$$

Final stage:

$$\begin{aligned} f_5(\text{A}) &= \min[(7 + f_4(\text{B})), (3 + f_4(\text{C}), 2 + f_4(\text{D}))] \\ &= \min[7 + 12, 3 + 14, 2 + 22] = 17 : \text{ACEILN} \end{aligned}$$

Therefore the minimum distance is 17, and the route to follow is ACEILN. The reader should now attempt to repeat the exercise but this time adopt the notation that $f_j(\beta)$ is the minimum distance *from* A *to* point β over j stages. Naturally, the identical minimum-distance results should obtain!

7.4 APPLICATION TO THE SELECTION OF EQUIPMENT

A problem which regularly appears in optimization is that of maximizing (or minimizing) some function subject to limited resources where the value of the function depends upon whatever resources are available. The following is a reasonable example of this type of allocation problem. A space capsule is to be loaded with a number of experiments of which there are four types. The weight of each experiment and the value to scientists in terms of mission success is known. The problem is to choose how many of each type of experiment to load into the capsule so as to maximize the value of the payload subject to a given weight limit. (N.B. Duplication of experiments or experimental packages implies greater reliability in the mission and increased confidence of the scientific data being collected.) The data for this problem are shown in Table 7.1. It should be assumed that both the weight and value functions are proportional to the number of experiments of their appropriate type, thus giving a linear problem but with integral constraints (the problem could be solved using integer linear progamming (ILP) algorithms).

We start as we always do by posing the question, 'since it is a maximum value we seek, upon what will this maximum value depend?' Let the maximum value be denoted by f, then f must depend on the weight limit (the more experiments we are allowed to load the greater will be the

Table 7.1

Experiment type	Value (v_i)	Weight (W_i)
1	5	2
2	8	3
3	14	5
4	11	4

maximum value) and also upon the number of experiment types from which we have to choose. In our problem these values are 9 and 4 but let us say that in the general case they may be denoted by W and N respectively. Therefore:

$$f = f_N(W)$$

What then is $f_1(W)$?

We denote $[y]$ to be the largest integer smaller than y, e.g. $[7.5] = 7$. Then clearly

$$f_1(W) = [W/W1] \cdot v1$$

and more precisely

$$f_1(W) = [W/2] \cdot 5$$

We can therefore tabulate $f_1(W)$ for all possible values of weight limit which we may need to consider in the range 0 to 9 inclusive, thereby producing Table 7.2.

The decision of how many experiments of type 1 to load is of course denoted by x_1. What then of $f_2(W)$? Our initial decision, let it be x_2, does not matter so far as the principle of optimality is concerned. But the value to be obtained from the remaining decision should be, indeed must be a maximum with respect to the state resulting from that first decision. That state will clearly be $W - w_2x_2$. The total value will be:

$$v_2x_2 + f_1(W - w_2x_2)$$

an expression that depends only on x_2, the initial decision. If this is now chosen so as to maximize the expression, it must yield the appropriate value of $f_2(W)$, wherefore

Table 7.2

W	$f_1(W)$	x_1
0	0	0
1	0	0
2	5	1
3	5	1
4	10	2
5	10	2
6	15	3
7	15	3
8	20	4
9	20	4

$$f_2(W) = \text{maximum } \{v_2x_2 + f_1(W - W_2x_2)\};\ 0 < x < [W/W_2]$$

This reasoning is easily extended to the general case, whereupon the recurrence relations for the entire problem because:

$$f_N(W) = \text{maximum}\{v_Nx_N + f_{n-1}(W - W_Nx_N)\};\ 0 < x_N < [W/W_N]$$
$$N = 2, 3, \ldots$$
$$f_1(W) = [W/W_1] \cdot v_1$$

In detail the relations for our particular problem are:

$$f_1(W) = [W/2] \cdot 5$$
$$f_2(W) = \text{maximum}\{8x_2 + f_1(W - 3x_2)\};\ 0 < x_2 < [W/3]$$
$$f_3(W) = \text{maximum}\{14x_3 + f_2(W - 5x_3)\};\ 0 < x_3 < [W/5]$$
$$f_4(W) = \text{maximum}\{11x_4 + f_3(W - 4x_4)\};\ 0 < x_4 < [W/4]$$

It therefore remains only to tabulate the $f_K(W)$ and x_k for $k = 2$, 3 and 4 against all possible values of W in the range 0, 9. This follows now in suitable stages.

Second stage (Table 7.3):
For $W = 0$, 1 and 2, $x_2 = 0$ only, therefore

$$f_2(W) = f_1(W)$$
$$f_2(0) = 0,\ f_2(1) = 0,\ f_2(2) = 5$$

For $W = 3$, 4, or 5, x_2 may be either 0 or 1, therefore

$$f_2(W) = \max[f_1(W),\ 8 + f_1(W - 3)]$$
$$f_2(3) = \max[f_1(3),\ 8 + f_1(0)] = 8 \text{ and } x_2 = 1$$
$$f_2(4) = \max[f_1(4),\ 8 + f_1(1)] = 10 \text{ and } x_2 = 0$$
$$f_2(5) = \max[f_1(5),\ 8 + f_1(2)] = 13 \text{ and } x_2 = 1$$

Table 7.3

W	$f_2(W)$	x_2
0	0	0
1	0	0
2	5	0
3	8	1
4	10	0
5	13	1
6	16	2
7	18	1
8	21	2
9	24	3

For $W = 6$, 7 or 8, x_2 may be either 0, 1 or 2 and therefore:

$$f_2(W) = \max[f_1(W), 8 + f_1(W - 3), 16 + f_1(W - 6)]$$
$$f_2(6) = \max[15, 13, 16] = 16 \text{ and } x_2 = 2$$
$$f_2(7) = \max[15, 18, 16] = 18 \text{ and } x_2 = 1$$
$$f_2(5) = \max[20, 18, 21] = 21 \text{ and } x_2 = 2$$

For $W = 9$, x_2 may be 0, 1, 2 or 3 and therefore:

$$f_2(9) = \max[20, 23, 21, 24] = 24 \text{ and } x_2 = 3$$

Third stage (Table 7.4):
For $W = 0$, 1, 2, 3, or 4, x_3 may only be 0, therefore

$$f_3(W) = f_2(W)$$

For $W = 5$, 6, 7, 8 or 9, x_3 may be 0 or 1, therefore

$$f_3(W) = \max[f_2(W), 14 + f_2(W - 5)]$$
$$f_3(5) = \max[f_2(W), 14 + f_2(0)] = 14 \text{ and } x_3 = 1$$
$$f_3(6) = \max[f_2(W), 14 + f_2(1)] = 16 \text{ and } x_3 = 0$$
$$f_3(7) = \max[f_2(W), 14 + f_2(2)] = 19 \text{ and } x_3 = 1$$
$$f_3(8) = \max[f_2(W), 14 + f_2(3)] = 22 \text{ and } x_3 = 1$$
$$f_3(9) = \max[f_2(W), 14 + f_2(4)] = 24 \text{ and } x_3 = 0 \text{ or } 1$$

Table 7.4

W	$f_3(W)$	x_3
0	0	0
1	0	0
2	5	0
3	8	0
4	10	0
5	14	1
6	16	0
7	19	1
8	22	1
9	24	0 or 1

Final stage (Table 7.5):
For $W = 0$, 1, 2 or 3, x_4 may only be 0, therefore

$$f_4(W) = f_3(W)$$

For $W = 4$, 5, 6 or 7, x_4 may be either 0 or 1, therefore

$$f_4(W) = \max[f_3(W), 14 + f_3(W - 4)]$$

Table 7.5

W	$f_4(W)$	x_4
0	0	0
1	0	0
2	5	0
3	8	0
4	11	1
5	14	0
6	16	0/1
7	19	0/1
8	22	0/2
9	25	1

$$f_4(4) = \max[10, 11 + 0] = 11 \text{ and } x_4 = 1$$
$$f_4(5) = \max[14, 11] = 14 \text{ and } x_4 = 0$$
$$f_4(6) = \max[16, 11 + 5] = 16 \text{ and } x_4 = 0 \text{ or } 1$$
$$f_4(7) = \max[19, 11 + 8] = 19 \text{ and } x_4 = 0 \text{ or } 1$$

For $W = 8$ or 9, x_4 may be 0 or 2, therefore

$$f_4(8) = \max[22, 11 + 10, 22 + 0] = 22 \text{ and } x4 = 0 \text{ or } 2$$
$$f_4(9) = \max[24, 11 + 14, 22 + 0] = 25 \text{ and } x_4 = 1$$

The result we seek, $f_4(9)$, is therefore 25. How we obtain this is to load one of type 4, weight = 4 and value = 11, leaving room for a weight of 5. From Table 7.4 we see that $f_3(4) = 14$ and $x_3 = 1$. Therefore the optimal combination is to pack one each of type 3 and type 4. Note, however, that our computational effort will now enable use to solve any problem for $N \leqslant 4$ and with any weight limit of $W \leqslant 9$.

For example, for $N = 3$ and $W = 5$, the solution is: $x_1 = 0$, $x_2 = 0$, $x_3 = 1$, $x_4 = 0$, and the maximum value is 14; and for $N = 3$ and $W = 8$, the solution is $x_1 = 0$, $x_2 = 1$, $x_3 = 1$ and the maximum value is 22.

7.5 FURTHER NETWORK PROBLEMS AND HEURISTIC METHODS

This section underlines the prevalence of network-type problems which the systems engineer encounters and the power of network models to resolve these problems. An interesting development in network modeling is the use of what are called heuristic methods for obtaining attractive, near-optimal solutions. While not always being provably optimal, these are useful to the systems engineer since they incorporate a pragmatic element to their solution. In the next chapter, a more detailed treatment of heuristic methods will be given under the wider heading of artificial intelligence. For the moment, the reader is encouraged to explore the subtlety of approach that is found in a simple treatment of those problems for which a network model is thought fitting. It is important to realize that networks are a form of model, very significant in many respects, for a large number of real-world situations.

A network is conveniently thought of in terms of two sets: nodes and links. The links are a collection of individual associations between two nodes. The nodes may represent, purely for convenience, meeting points of the links, having no real meaning of their own, e.g. the end of a set of links, the beginning of another set, or they may represent physical space where activities associated with the links, e.g. traffic or power movement continue. In this situation not only do the nodes have a physical meaning of their own but may well dominate the entire behavior of the network. In general both nodes and links have active and passive parts to play in the constantly changing behavior of the network. If these basic factors of the network cannot be appreciated now, the understanding of even a physical situation seen as a network may be lost. In other words, physical situations are what they are; they may become networks when they are seen as such. Their value of being seen as networks can only be fully appreciated if the physical parts which become either the nodes or the links can be properly appreciated in this modeling process.

To illustrate, the following are some physical situations which may be, and indeed have been, regarded as suitable for consideration as network problems:

1. Transportation of goods between suppliers, warehouses and retail outlets.
2. Network planning for the transmission and distribution of electricity, gas, oil and water.
3. Communication between individuals, message stations and message centers by letter, telegraph, telephone and data links.
4. Road and rail systems with applications in motorway construction, road-rail links, timetabling and allocation of resources.

In each of the above a host of network problems can be envisaged. The following section concentrates on the analysis and solution of three principal network problems in addition to providing a précis of the state-of-the-art of network algorithms. The following particular algorithms and related problems are discussed:

1. Shortest route problem and auxiliaries.
2. Maximum capacity route problem.
3. Maximum flow problems.

7.5.1 Networks: models and algorithms

The first type of network model to consider is the shortest-route problem and its various auxiliaries. The generic model is stated as follows: assign distance values to all the links in the network and find the path from i to j such that the sum of all the distances of links traversed is a minimum. There may be several shortest routes or minimum-distance paths. It may be necessary to do this process for all pairs if the problem requires it.

There are several ways to solve this problem and most of them make use of the fact that the optimal (minimum-distance) route contains sub-routes that are also optimal. However, combining two optimal routes may not produce an optimal route. In other words in the extension of an

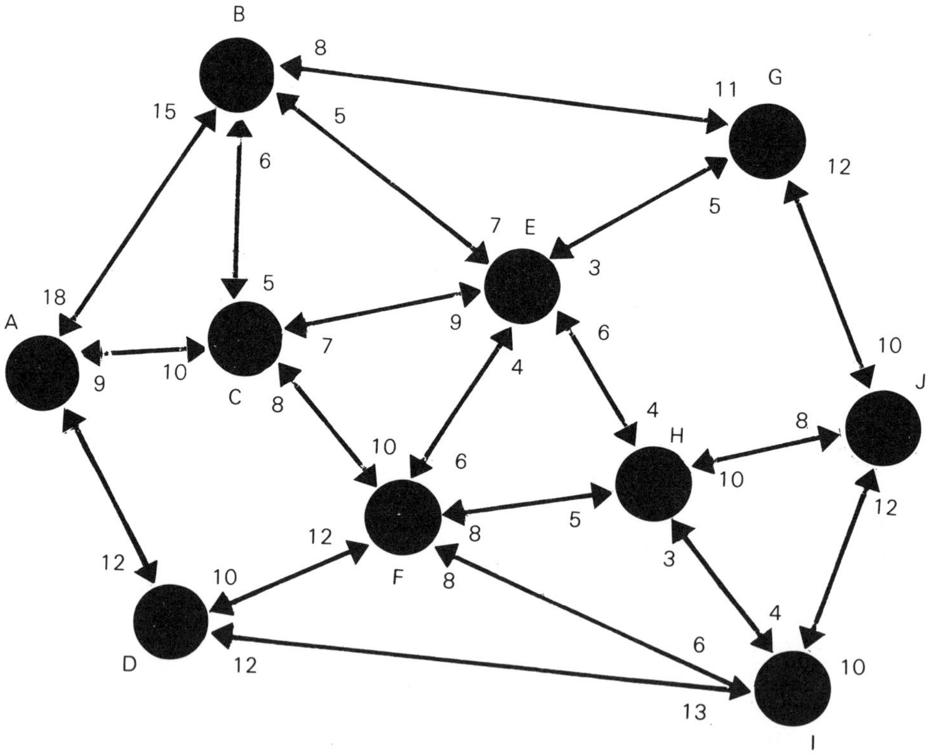

Figure 7.3

optimal route which automatically becomes and optimal subroute to form a new route, if the new route is to be optimal the extension must be carried out over all possible summations of optimal subroutes with their extensions.

This is illustrated by the labeling algorithm which is presented below and applied to the network in Figure 7.3. The method is due to Minty (Carre, 1980). Let us suppose that at some stage we have found the shortest routes to r of the nodes from a specified origin node. These nodes, x_i, $i = 1, 2, \ldots, r$, will be labeled with the shortest distance from the origin, $M(X_i)$. To find the next node and link in the extension of these routes, we list all the links that lead from a labeled node, being referred to as at the tail of the link, to an unlabeled node, at the head of the link, one link away. These nodes are called adjacent nodes, Y_j, $j = 1, 2, \ldots, k$, say. Let the length of the link from X_j to Y_j be $x_{ij} > 0$. For each of the links we form the sum of the length of the links and the label at its tail. Each of these sums represents the distance of an unlabeled node from the origin along a route which is at least optimal as far as the last but one node. The nodes and link for which the sum is smallest are suitably marked and labeled, becoming part of the extending shortest routes.

The proof of this is simple. Let the i, j for the minimization be p and q so that

$$m(X_p) + x_{pq} < m(X_i) + x_{ij} \forall i, j \tag{7.1}$$

Any other path to Y_q that has not been taken into account by the above must have at least two links between an X_j and Y_q, one of which must be from an X_j to some Y_q along such a path must be of the form

$$m(X_i) + x_{ij} + y, \qquad \text{where } y > 0 \tag{7.2}$$

From (7.2) it follows that

$$m(X_i) + x_{ij} + y \geq m(X_p) + x_{pq} \quad \text{for} \quad i \neq p, j \neq q \tag{7.3}$$

The algorithm is:

Step 1: Label the origin node with 'distance' zero and identify the terminating node.

Step 2: Search out the links whose tail ends are labeled and whose heads are unlabeled.

Step 3: For each of the links found in Step 2 form the sum of the label at the tail end and its length.

Step 4: Mark the link which has the least of the sums in Step 3 and label the head of this link with this sum.

Step 5: If the terminating node has been labeled, the algorithm stops and the shortest route is that between origin and terminator consisting of marked links. Otherwise go to Step 2.

This algorithm was applied to the network above by hand computation. It should be noted that the shortest routes are those 'away' from node A since distances are not symmetric. Distance can therefore be generalized to mean cost, time or capacity.

Because there are a variety of methods to solve this problem, no one method is the best and that includes this one. The *k* best routes problem is a generalization of the optimal route problem. For a distance network it would be to find the $(k - 1)$ shortest routes, between a particular pair or pairs of nodes, after the shortest route.

The second type of general network model is the maximum capacity route which can be stated thus: given a set of nodes and links, each link having a non-negative capacity c_{ij} ($= 0$ if i and j do not connect directly), find the route from node n to m that allows the maximum flow to take place. For any given route the capacity will be equal to that of the link whose capacity is the minimum among the links on that route.

Shortest route methods can be modified to cover this problem. For example, Minty's algorithm would now read as shown below.

Step 1: Label origin with capacity ∞. (All nodes are assumed to have an infinite capacity initially.)
Step 2: Search out the links whose tail ends are labeled and whose heads are unlabeled.
Step 3: For each of the links found in Step 2 form the minimum of the label and the link capacity.
Step 4: Mark the link which has the maximum capacity of the minima found in Step 3 and label the head of this link with this value.
Step 5: Stop if the terminating node is labeled, the maximum capacity route having been found (and marked). Otherwise go to Step 2.

For what it is worth, and we will see just exactly what later in the test, a minimum capacity route is easily calculated by having Step 4 as: 'Mark the link which has the minimum capacity of the minima found in Step 3 and label the head of this link with this value.'

The final type of general network model is the maximum flow problem. This problem differs from the maximum capacity route problem in that we wish to determine the maximum flow possible between nodes m and n, say, making use of any of the several routes between them. More generally we might want to find the maximum flow possible between one set of nodes (the sources) and another set of nodes (the sinks). However, this problem is the same as finding the maximum flow between two nodes since we can define a new node M connected by infinite capacity links to the sources and a new node N connected by infinite capacity links from the sinks. The problem then is to find the maximum flow between M and N.

Each link in the network may be labeled with two numbers (f_{ij}, r_{ij}) where f_{ij} is the flow in the link (i, j) and r_{ij} is the residual with respect to the capacity C_{ij}. Thus:

$$f_{ij} + r_{ij} = C_{ij}$$

The algorithm for maximal flow, due to Ford and Fulkerson (Smith, 1982), also uses a labeling method where nodes are given two labels $(s_{j,i})$ where node j is given a capacity value s_j found from $s_j = \min(r_{ij}, s_i)$, node i being an adjacent node and also the second of the two labels. The algorithm is as follows:

1. Set up an initial flow pattern. (This may be all zeros or an existing flow pattern, or one thought to be near the expected optimum.)
2. Label source node with source capacity, S_c.
3. Label all unlabeled nodes j which are adjacent to a labeled node i with $(s_{j,i})$. Note that node j cannot be labeled if either there is no link (i,j) or the link has a zero residual. Stop labeling when either the sink is labeled, in which case go to Step 4, or no more labels can be assigned, in which case the algorithm stops and the flow pattern is the optimum.
4. When the sink node has been labeled (S_{sink}, i), the flow in the network can be increased by an amount S_{sink}, so far as those links are concerned which form the route from source to sink indicated by the ordered set of the second of the two node labels. Residuals are adjusted for each link accordingly.
5. When the extra flow has been accommodated, all the node labels are deleted and the algorithm proceeds from Step 2.

The reader should now consider the application of this algorithm to the network shown in Figure 7.4.

7.5.2 The method of branch-and-bound

We proceed now to consider a method which is becoming a standard against which all other heuristic searches of network models are judged. We introduce it by referring to a network-type problem of equally classical proportions which finds many analogies to problems routinely

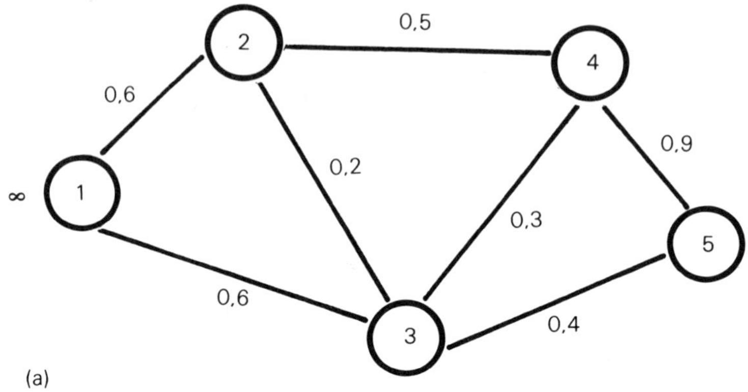
0,5
2
0,6
4
0,9
0,2
∞
1
0,3
5
0,6
0,4
3

(a)

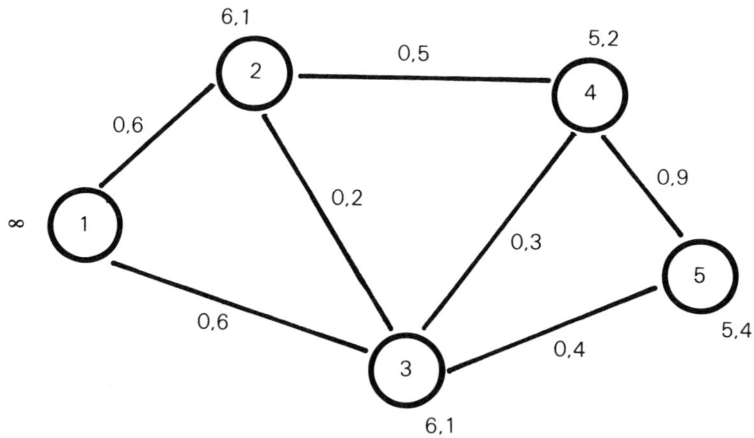
6,1
5,2
0,5
2
4
0,6
0,9
0,2
∞
1
0,3
5
0,6
5,4
0,4
3
6,1

(b)

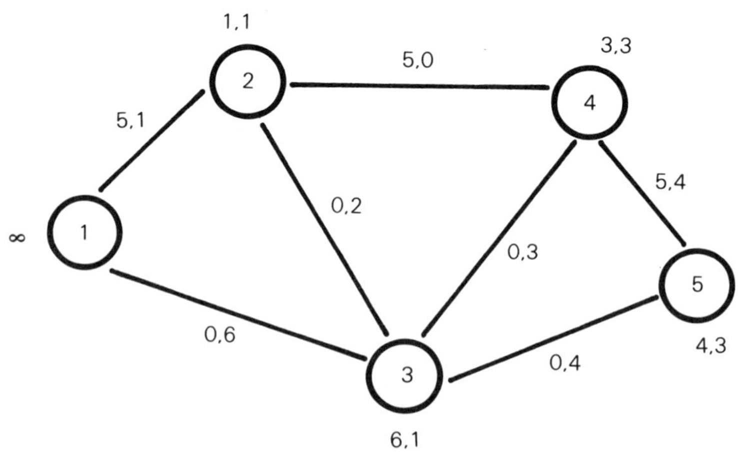
1,1
3,3
5,0
2
4
5,1
5,4
0,2
∞
1
0,3
5
0,6
4,3
0,4
3
6,1

(c)

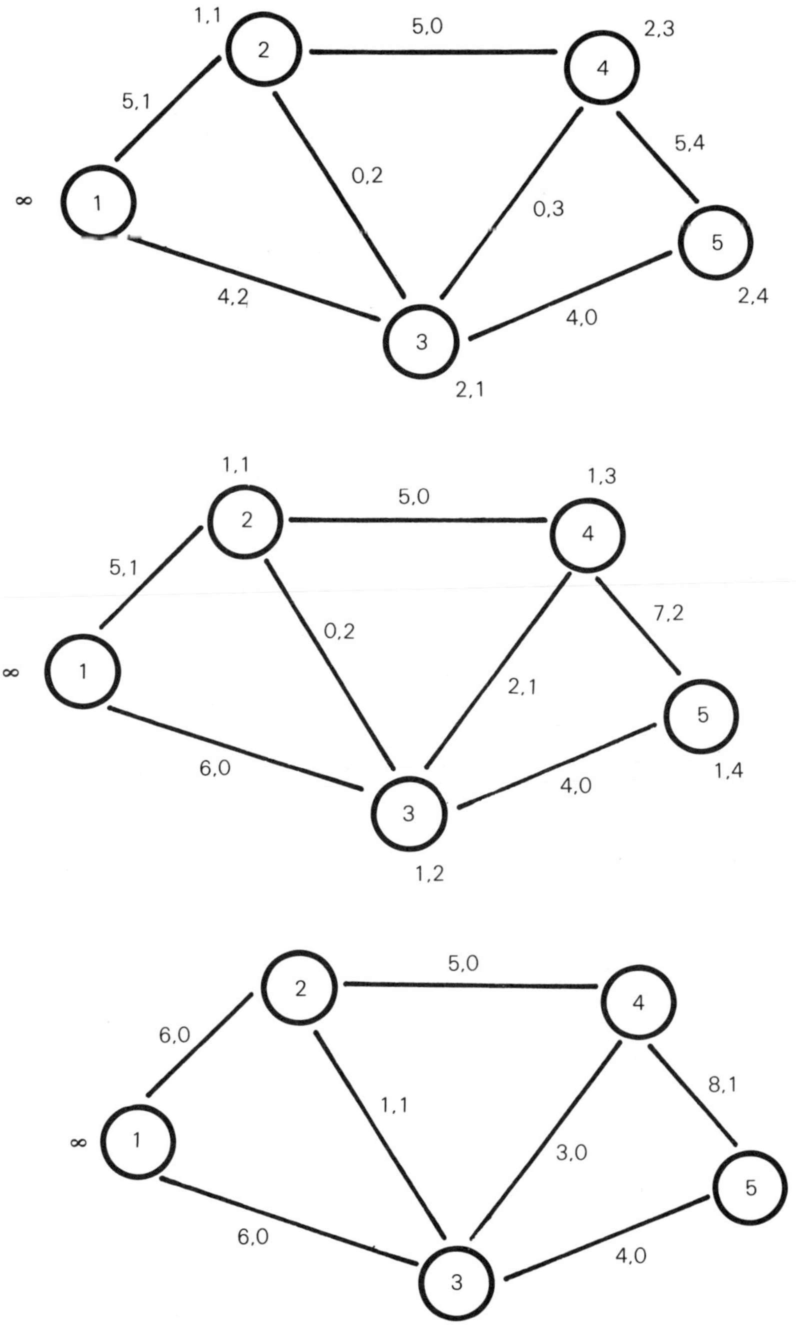

(d)

(e)

(f)

Figure 7.4

encountered by systems engineers. The method is branch-and-bound, and the problem is known as the traveling salesman problem. It is generally stated along the following lines. A salesman located in a given city must visit each of $N-1$ cities once and only once and then return to the city he started from. The question is posed: 'In what sequence should he visit the cities in order to minimize the total distance traveled?' This is a problem which is simple to state and is common in an industrialized society, and then not always in the guise of an itinerant peddlar! Notwithstanding the enormous effort that has been put into the solution of this problem, algorithms that find the exact optimal solution are still limited to fewer than 100 cities. Smith (1982) describes an algorithm that obtains a 'good' solution for problems with fewer than 250 cities in a reasonable amount of time on the fastest computers. However, algorithms that will find the exact solution to 'large' problems are still very much in demand.

The traveling salesman problem is in fact a simplified version of the more general vehicle scheduling problem. In this, a quantity of a given resource must be left at each city with the added constraint that the total demand over all cities exceeds the capacity for any one delivery so that several visits may be required back to base before the overall tour can be completed. This problem can be made more complicated in many ways by, for example, having: a fleet of vehicles of varying capacities; mixed resources; collection as well as delivery; specified delivery times, and so on. At this stage we reach a more lifelike situation than the initial network model gives room to suggest. This particular sort of problem does seem to be a subject more for operational researchers than for systems engineers. But in the design of physical networks such as telephone systems and electricity supply systems, the combination of the routing, replacement and allocation functions of the design process, in total, gives rise to situations where, generally speaking, networks models apply and often, quite specifically, the network algorithms developed for traditional operations research use may be gainfully employed.

For example, another standard problem for the operational researcher which has been fashioned to suit engineering network problems is the transportation problem. Suppose that a certain commodity is produced at a number of different plants and that this must be shipped to several holding points or warehouses for later distribution to stores, i.e. the commodity is produced at m plants for shipment to n warehouses. If it costs a certain amount to ship 1 unit from a given plant to a given warehouse; if each plant has a certain production capacity; and if each warehouse has a given requirement, then the transportation problem is to determine the number of units to be shipped from each plant to each warehouse in order to minimize the overall shipping cost. The superficial resemblance that this problem bears to, say, the design of an electricity supply system, has been sufficient in the past to encourage and inspire not a few researchers to build just one or two bridges between this field of optimization and their own subject area of electrical engineering science.

Returning to the traveling salesman problem, it is possible to formulate this as an integer programming problem, but again its solution using this approach would be very expensive in computer time. In some references, there is a suggestion of an approach that amounts to an ingenious way of shortening the computation involved in a full integer programming method. Additionally, others have suggested a dynamic programming formulation to the problem but, again, for the larger-sized problem, this is expensive in terms of both computing time and storage requirements. The most promising new approach which has a wide range of applications to traditional and original network problems is the branch-and-bound method. It is computationally pragmatic but requires a thorough understanding of both the problem and of branch-and-bound formalism in order to achieve solutions of an acceptable quality. That there have been so many different, inventive and relatively successful algorithms put forward must suggest the need for this type of work. Indeed once we are satisfied that this need is valid, our next point must surely be to concentrate on the fact that there are so many different ways of dealing with any given problem. The abundance of alternatives fixes one's attention on the merit of network algorithm development as a ruling principle in the education and experience of the systems engineer.

We now turn our attention to the various algorithms which have been devised to solve, in different forms, the traveling salesman problem, and in so doing we encounter the branch-and-bound technique. This is generally described as a well-structured systematic search of the space of all feasible solutions of constrained optimization problems that have a finite number of feasible solutions. Usually the space of all feasible solutions is repeatedly partitioned into smaller and smaller subsets (the branching process), and a lower bound (for a minimization problem) is calculated for the cost of solutions within each subset (the bounding process). After each partitioning, those subsets with a bound exceeding that of a known feasible solution are excluded from all further branching. Thus, large subsets of solutions may be excluded from consideration without examining each solution in these subsets. Branching continues until a known feasible solution is found with a bound no greater than that for any subset, at which point this is taken as the optimal solution. The success of the technique depends upon the number of solutions examined before an optimal solution is reached. It has, however, been applied with a great deal of success to the wide range of traditional network-type problems and therefore should fall within the attention of the systems engineer.

It is important to stress that like dynamic programming, the formulation of a branch-and-bound algorithm may vary considerably from problem to problem, supporting many diverse branching and bounding policies in turn. However, the ruling principle of the technique remains constant: at any point in a branch-and-bound algorithm where a branching decision must be made, any subset with a bound that is less than the least upper bound for all feasible solutions discovered to date is a possible candidate for branching (minimization problem).

Two basic branching decision rules are available:

1. Branch from lowest bound.
2. Branch from newest active bound.

The meaning of 1 above is that the next branching should be from the subset of possible solutions that has the lowest bound in the optimal solution. In practice this has the advantage that few subproblems are examined but the disadvantages that it requires more computer storage, because more intermediate data must be stored. The meaning of 2 is that the choice for branching is from the subsets most recently generated and that usually the lowest bound is the next branch.

Conversely rule 2 normally requires many more branching decisions to be made, but is far less demanding of computer storage. A branch-and-bound algorithm for the traveling salesman problem was proposed in Carre (1980). This requires, however, the intermediate solution of several problems all of the same kind, namely the assignment problem. It is helpful therefore before considering the algorithm for the traveling salesman problem, to look briefly at the assignment problem and how this itself may be solved using the branch-and-bound technique.

Suppose that a project involves m tasks and that at least m operatives are available to carry out the tasks. Any operative is capable of performing each task but the problem is to assign each operative a single task so that the time to perform all tasks overall is a minimum. Given the effectiveness matrix, relating the time that each operative takes over each task, the assignment problem is to determine which operative is assigned to which task.

This is just one example of the classical assignment problem for which several different algorithms and optimization methods have been proposed with varying degrees of success. A branch-and-bound algorithm for this problem is now described by direct application to an elementary

Table 7.6

Operatives	Task 1	Task 2	Task 3	Task 4
A	6	5	8	3
B	10	5	4	15
C	13	7	2	11
D	13	9	7	10

example involving four operatives and four tasks, the effectiveness matrix being shown in Table 7.6.

The branch-and-bound technique is in effect the construction of a tree network consisting of nodes and branches. It starts from a root node which represents the set of all possible solutions and whose bound is the least of all nodes. Branching from the root involves making a decision and arriving at another node, one of many possible, which represents a subset of solutions. The bounding policy is designed to exploit the particular situation to which the technique is being applied. For a minimization problem the lower bound is formed usually by ignoring the constraints which, when applied, distinguish feasible and infeasible solutions. Infeasible solutions can always be explored further by branching and the least upper bound that determines which partial solutions are to be further explored and which are not. It also provides the stopping criterion and hence the optimal solution. For our problem, indicated by the data of Table 7.6, the root is denoted by node 1 and has a lower bound of 16. This is arrived at by calculating the minimum time to do all four tasks ignoring the constraint that 'one and only one operative is assigned to each task'. In other words, the assignment:

$$A \rightarrow 1 \quad B \rightarrow 2 \quad C \rightarrow 3 \quad D \rightarrow 4$$

is the least of all bounds and represented by the root node of our branch-and-bound solution, Figure 7.5(a).

Our decision policy is to choose for each stage which of the four operatives should be assigned to the 'next' task. Our first decision is to determine which of the operatives should be assigned to task 1. The choice is clearly between four operatives and therefore four branches emanate from the root. The bound for each of nodes 2 to 5 inclusive are calculated on the basis of minimum-time to complete the remainder of the tasks, ignoring our chief constraint. These calculations are shown in Table 7.7 and the result shown in Figure 7.5(b). The least upper bound is 23 (nodes 2 and 5). Note that node 4 will never be further explored. We must explore node 3, however, since this is the current least bound on

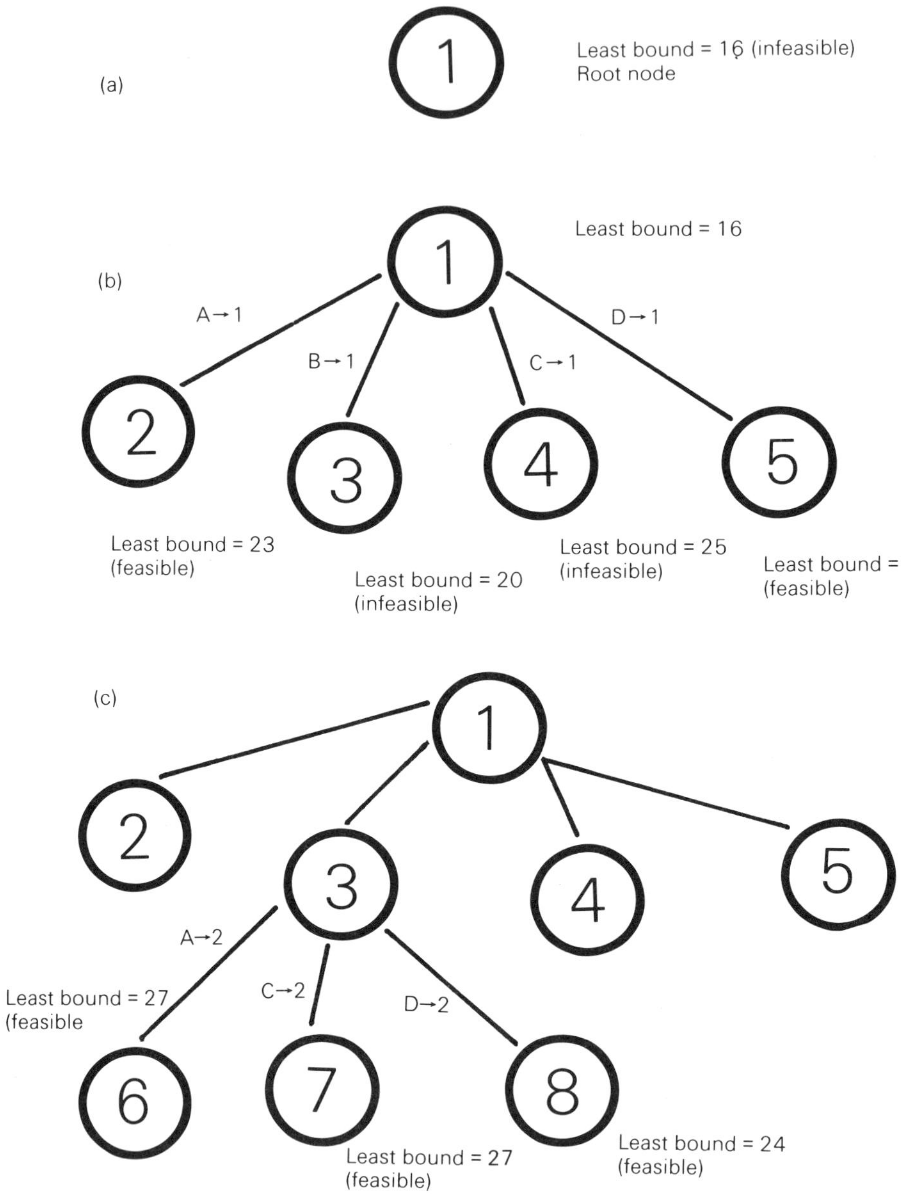
(a)
1
Least bound = 16 (infeasible)
Root node
(b)
1
Least bound = 16
A→1
B→1
C→1
D→1
2
3
4
5
Least bound = 23
(feasible)
Least bound = 20
(infeasible)
Least bound = 25
(infeasible)
Least bound = 23
(feasible)
(c)
1
2
3
4
5
A→2
C→2
D→2
Least bound = 27
(feasible
6
7
8
Least bound = 27
(feasible)
Least bound = 24
(feasible)

Figure 7.5

unexplored solutions and so we branch from these nodes, noting that this determines B to have been assigned to task 1.

The result of this branching is shown in Figure 7.5(c), where the least bounds have been calculated from Table 7.8. Clearly this second-stage branching has involved us in making a decision as to which of the remaining operatives is to be assigned to task 2. From the result of Figure 7.5(c)

Table 7.7 First-stage branching

	Task 2	Task 3	Task 4
B	5	4	15
C	7	2	11
D	9	7	10
Min.	5	2	10 = 17
Operative	B	C	D

(a)

	Task 2	Task 3	Task 4
A	5	8	3
C	7	2	11
D	9	7	10
Min.	5	2	3 = 10
Operative	A	C	A

(b)

	Task 2	Task 3	Task 4
A	5	8	13
B	5	4	15
D	9	7	10
Min.	5	2	3 = 12
Operative	A	B	A

(c)

	Task 2	Task 3	Task 4
A	5	8	3
B	5	4	15
C	7	2	11
Min.	5	2	3 = 10
Operative	B	C	A

(d)

Table 7.8 Second-stage branching

	Task 3	Task 4
C	2	11
D	7	10
Min.	2	10 = 12
Operative	C	D

(e)

	Task 3	Task 4
A	8	3
D	7	10
Min.	7	3 = 10
Operative	D	A

(f)

	Task 3	Task 4
A	8	3
C	2	11
Min.	2	3 = 5
Operative	C	A

(g)

we see that the current least upper bound is 23 (nodes 2 and 5) and that this is the least bound on all solutions. The optimal solution has been found therefore and results in either of the two assignments:

$$\begin{array}{llll} A \to 1 & B \to 2 & C \to 3 & D \to 4 \\ A \to 4 & B \to 2 & C \to 3 & D \to 1. \end{array}$$

The reader should attempt exercise 14 as an aid to developing confidence in the branch-and-bound technique.

That slight digression from our main subject of the traveling salesman problem will serve us in good stead as we now turn to the branch-and-bound algorithm proposed by Eastman (Carre, 1980) to solve the same.

In the following it is understood that a 'tour' is a complete route through the N cities where no city is visited more than once. If the salesman visits a certain city and returns to that city later, the cities involved form a 'subtour'. This cannot occur if a route is feasible (each city is visited once and only once). Eastman's algorithm solves the easier assignment problem that allows subtours and then systematically forbids subtours until finally a tour is obtained that is optimal:

Step 1: Let x represent the current least upper bound on the optimal solution and set $x = \infty$.

Step 2: Solve the associated assignment problem, where the distances d_{ij}, are the elements of the effectiveness matrix. This solution provides a lower bound on the optimal solution. If at least one subtour exists in the solutions, go to Step 3. Otherwise the optimal solution of the assignment problem is also an optimal solution of the traveling salesman problem, so stop.

Step 3: Select the subtour with the smallest number of links and ignore all others at this node. (N.B. Branching must be done for only one subtour at each node). Suppose that this subtour has k links and can be denoted as:

$$l_1 \to l_2 \to \ldots \to l_k \to l_1$$

where l_i are the cities on the subtour.

Step 4: Branch into k subproblems and for the tth subproblem:

Step 5: Solve the k new assignment problems. Each solution distance is a lower bound for the corresponding problem.

Step 6: If there are one or more feasible solutions from Step 5, then let the smallest total distance of these be Y. If Y is smaller than X, then let X become equal to Y and solve the corresponding feasible solution. Otherwise X remains unchanged.

Step 7: If X is the least bound then the corresponding solution is an optimal solution of the traveling salesman problem, so stop. Otherwise go to Step 8.

Step 8: From the set of all unexplored non-feasible (i.e. subtours present) subproblems with a bound less than X, select the subproblem with the least lower bound for further branching. Go to Step 3.

The reader is referred to exercise 15 to have an opportunity to test out this algorithm.

7.6 NETWORK ANALYSIS APPLIED TO PROJECT PLANNING

7.6.1 The problem

Not only can network models be devised of systems where transportation of physical resources is involved, but also the same network modeling approach can be used equally effectively in the management of projects where time is of the essence. The name given to the latter is critical path planning, where network models of large and complex projects can be manufactured and monitored so as to give effective control of the overall project.

There are many questions asked by management before undertaking such projects, since the allocation of resources and a prior knowledge of performances, breakdown and general control, where possibly millions of pounds (£) is vitally important in safeguarding investment. We begin with some preliminary observations. A project is a large complex of individual jobs. Some of these can be started at the beginning of the project, others must wait until previous jobs have been completed. From previous experience a manager will have a good idea of how long individual jobs will take. How can he best use this knowledge to solve questions similar to the following?

1. How long will the project take?
2. At which points in time can the project controller expect to start and finish various jobs?
3. If the project has a scheduled completion date, what are the latest points in time various jobs can be started and finished subject to the project being finished by no later than this date?

4. How long can any job be delayed without affecting the scheduled project date or project duration?
5. If a particular job takes half (or twice) as long as it was originally estimated to take, how will this affect the project duration?

7.6.2 Critical path analysis

Basic concepts

Critical path analysis (CPA) offers a means of obtaining questions to the above, and so we begin with some basic concepts:

1. Activities, drawn as arrows, represent individual jobs. The tail of the arrow is the start of the job; the head is the end of that job. The job proceeds, naturally enough, from tail to head.
2. Events are points in time, taking no time, representing the start of some jobs and the finish of others. They are variously represented as circles, circles with crosses inside and circles with numbers inside, where these numbers represent the labels of the event.

These two concepts, the activity and the event, are put together in a way which reflects the logic of the project being modeled. Suppose activity B cannot start until A has finished. This could be represented by Figure 7.6 where B cannot start until 1 has happened, which will be when A is completed. Figure 7.7 shows a network model of a typical domestic situation where a young married couple get up in the morning. What is noticeable is that the entire model consists simply of activities (arrows), and events (circles) and nothing else. One special point about the model is that one activity is denoted by a broken arrow. This has a special name, the dummy activity, which in fact lasts no time at all but is introduced to make the logic of the model consistent with that of the project. In our case the wife does not make the tea until the kettle has been put on (and presumably boiled) *and* she has dressed. It might be suggested that the same result could be achieved by identifying events 13 and 16 as common. This would be so but it would produce the further implication that the wife could not cook breakfast until the kettle has boiled, which is quite unnecessary. Therefore the use of the dummy activity is of great

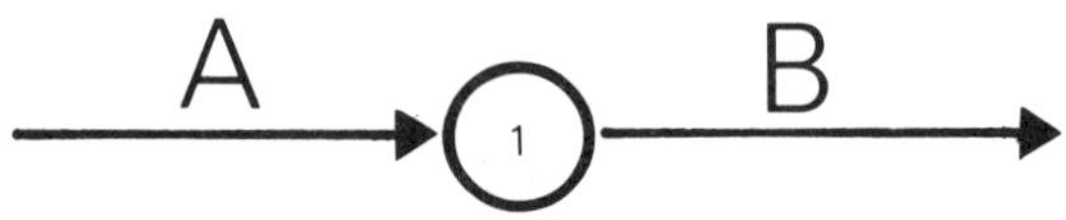

Figure 7.6

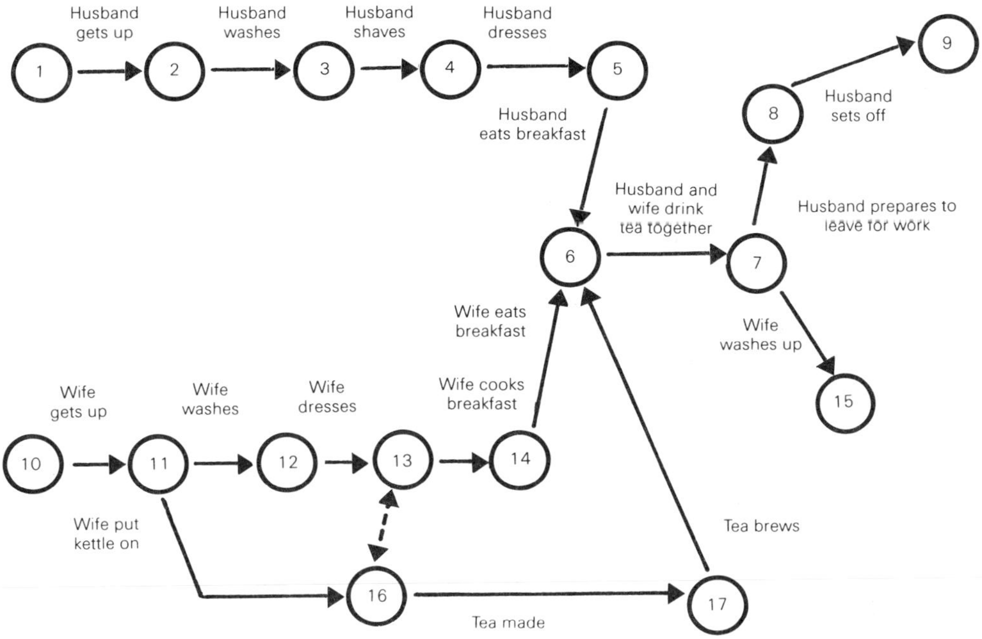

Figure 7.7

advantage. A further point with the network model is that it appears that the husband may eat his breakfast before it is actually cooked. Nowhere is there an expression of dependence of the former on the latter. A dummy activity (14, 5) would ensure this. Notice once again that it would be wrong to identify these two events as common since that would create a dependence (14, 6) on (4, 5), i.e. the wife could not eat her breakfast until it was cooked, and her husband had finished dressing – once again an unnecessary constraint!

If we complicate this domestic project a little more by insisting that there is only a single washbasin and that one or the other must take a turn at waiting, we shall see further the advantage of the dummy activity. Figure 7.8 shows a completed network model, dummy (14, 5) included, with the situation of the single washbasin. The reader is advised to ascertain which of the two, the husband or the wife, will always wash first by this model.

Style of analysis

Having discussed the network modeling of a project, the simulation exercise, involving the analysis of a network, must now be considered. There are five stages in the analysis of a network:

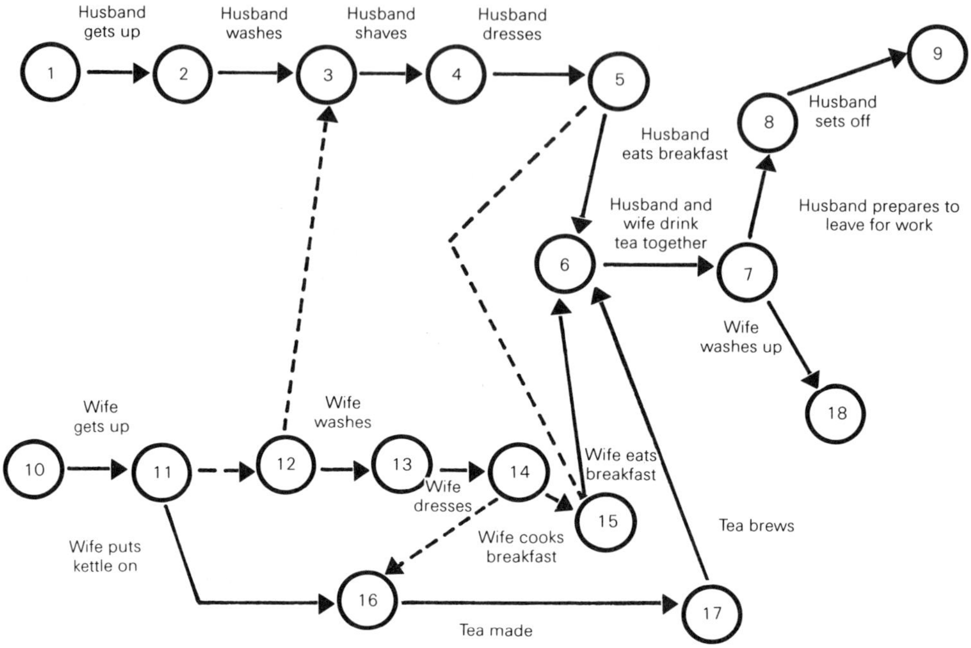

Figure 7.8

1. Calculate earliest or expected event dates.
2. Calculate latest event dates.
3. Calculate spare time at events and on activities.
4. Prepare a schedule.
5. Updating – due to changes in plan, actual progress in the network, etc.

Let us suppose we have a small project for which a network has been drawn up and agreed upon by all levels of personnel. The next exercise is to place figures against each activity to indicate their duration. Figure 7.9 shows a small example where the numbers represent the days. Let us assume that the project starts at time 0. Our first task is to establish the earliest event dates which effectively means the earliest time by which activities emanating from the events may start. As far as events 2 and 7 are concerned, these dates are obvious since activities (1, 2) and (1, 7) take four and five days respectively and we should expect the activities emanating from these events to commence immediately. Therefore these dates are 4 and 5 for events 2 and 7. However, for event 8, the earliest date is not immediately obvious. In fact, activity (8, 9) cannot start until

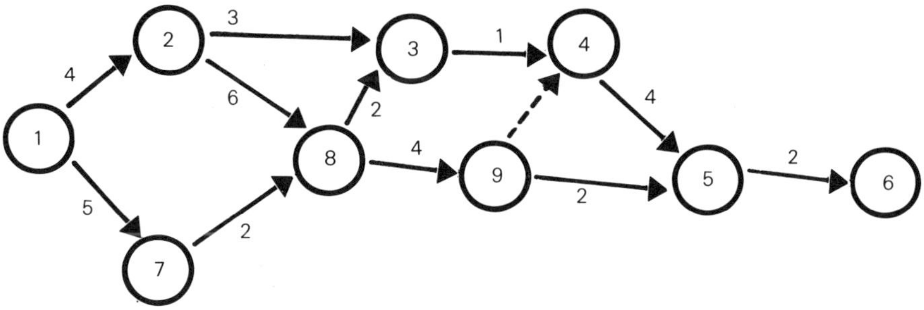

Figure 7.9

both (2, 8) *and* (7, 8) are complete. Therefore its earliest even date is the larger of the two dates 7 (= 5 + 2) and 10 (= 4 + 6), i.e. 10. This leads to the general rule: 'Whenever there is more than one activity terminating at an event, the largest sums of the durations of those activities and the earliest event dates of the events which precede each of them is taken as the earliest event date.'

Mathematically this becomes

$$E_j = \max\,(E_{il} + d(i_{l,j}),\ E_{i2} + d(i_{2,j}),\ \ldots$$

where $d(i_p, j)$ = expected duration of the event (i_p, j) for $p = 1, 2, \ldots, m$ where m is the number of activities terminating at event j. Figure 7.10 shows the earliest event dates for the small project being considered.

The second stage in the analysis is to calculate what are known as the latest event dates. Firstly, it is necessary to choose a scheduled completion date for the project, or in the general case, for each *end* event. This represents the latest point in time that the end event may occur. Then it is necessary to work backwards through the network, calculating for each event the latest date for its commencement if the overall project date is not to be delayed. In our case we assume 20 to be the project completion date. Then event 5 has a latest event date of 18. Event 4 has a latest event date of 14. What is the latest event date for event 9? It must not start later than 16 for (9, 5) takes two days and if 9 started later than 16 the project completion date would be delayed. But of more importance, the dummy takes no time and if event 4 does not happen by day 14, the project completion date would be delayed. Therefore event 9 must not happen later than 14. The general rule is: 'Whenever there is more than one activity starting from an event, the latest event date is the least of the differences between the latest event dates of events which immediately

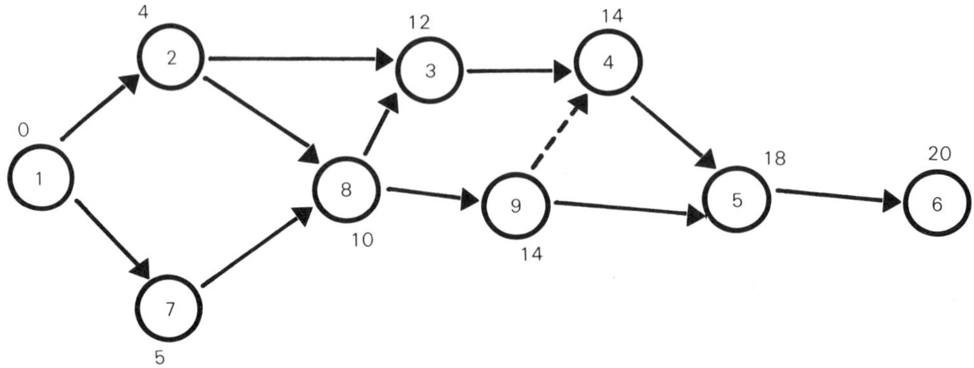

Figure 7.10

succeed the event in question and the duration of the activities starting from the event.'

Correspondingly, the mathematics is:

$$L_j = \min(L_{kl} - d(j,k_1)),\ L_{k2} - d(j,k_2),\ \ldots$$
$$L_{kn} = d(j,k_n)$$

where $d(j,k_p)$ = durations of activity (j,k_p), $p = 1, 2, \ldots, n$, where n is the number of activities starting from the event j. Proceeding back through the network in our case produces Figure 7.11.

The third stage in the analysis is the calculation of spare time at events and activities. An event cannot occur before its expected date, and should not occur after its latest date, but it can occur at any point in time

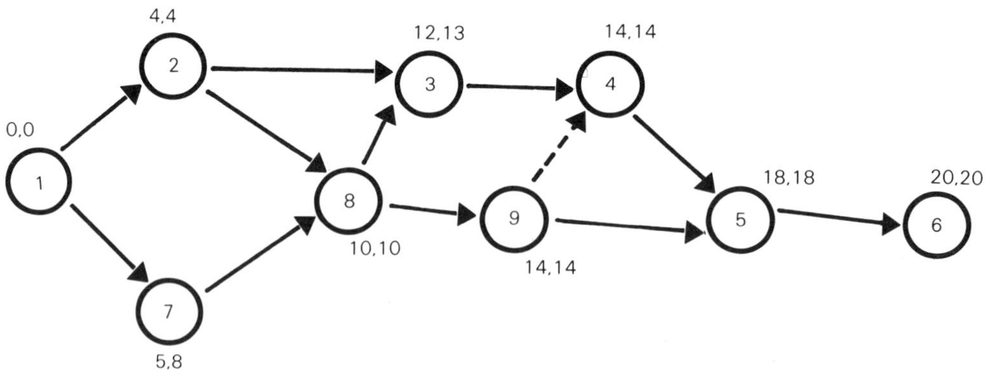

Figure 7.11

between these two. The difference between the expected and latest dates of an event is the slack if the event representing the amount of time the event could be delayed without affecting the scheduled completion date for this project. The events of zero slack form a path through the network which is called the critical path. No activity lying on the critical path may be delayed without a corresponding delay in the project. Such activities are referred to as critical activities.

The fourth stage of the critical path analysis is to prepare a schedule which is a list of the activities, showing when we *expect* to start and finish them and when we *must* start and finish them. There are five points which we would like to know about every activity:

1. The time it is expected to start (or the earliest it can start), the expected start E_s.
2. The time it is expected to finish or the earliest time it can finish, the expected finish E_f.
3. The time it must start or the latest start L_s.
4. The time it must finish, or latest finish, L_f.
5. The amount of time it could be delayed without affecting the scheduled completion date. This quantity is sometimes call the primary slack, P_s.

These quantities are calculated as follows:

1. E_s is obviously the expected date of activity's previous event.
2. $E_f = E_s +$ activity duration.
3. L_f is obviously the latest date of the succeeding event.
4. $L_s = L_f -$ activity duration.
5. $P_s = L_s - E_s = L_f - E_f$.

To prove the relationship of:

$$\begin{aligned} L_s - E_s &= L_f - d_{ij} - (E_s) \\ &= L_f - d_{ij} - (E_f - d_{ij}) \\ &= L_f - E_f \end{aligned}$$

If we define the span of an activity as the latest date of its succeeding event (L_f) minus the expected date of the preceding event (E_s), then:

$$\begin{aligned} \text{Span} &= L_f - E_s \\ &= (L_f - E_f) + (L_s - E_s) - (L_s - E_f) \\ &= L_f - E_f + (E_f - E_s) \end{aligned}$$

In other words:

Span − primary slack = duration

or

Primary slack = span − duration

and the span is the time available for the performance of an activity.
The schedule for our network is shown in Table 7.9.

Table 7.9

Activity	Duration	E_s	E_f	L_s	L_f	Slack	Span
1,2	4	0	4	0	4	0	4
1,7	5	0	5	3	8	3	8
2,3	3	4	7	10	13	6	9
2,8	6	4	10	4	10	0	6
3,4	1	12	13	13	14	1	2
4,5	4	14	18	14	18	0	4
5,6	2	18	20	18	20	0	2
7,8	2	5	7	8	10	3	5
8,3	2	10	12	11	13	1	3
8,9	4	10	14	10	14	0	4
9,4	0	14	14	14	14	0	0
9,5	2	14	16	16	18	2	4

7.6.3 Conclusions

In this second application of network analysis, to the planning and management of large-scale projects, we have further demonstrated the power of this modern design aid. Our discussion of project planning has been fairly general and of necessity rather elementary. There are, of course, several formally constituted processes, of which that described is but a simplified illustration, though interestingly they all prominently display a relationship to network analysis and, of greater significance, to systems engineering itself.

The program evaluation and review technique (Pert), which is a structured technique, basically similar to the process we have been discussing, but considerably more elaborate in detail, is an important management aid for expediting and controlling the utilization of personnel, materials, facilities and time consumed by the project.

It is used to pinpoint critical areas in a project so that necessary adjustments can be made in order to meet the scheduled completion date of the project. Pert is most applicable to large-scale research and development projects which involve a high degree of uncertainty, such as new project and development and marketing. Pert was developed in 1958/9 as a research and development tool for the United States Navy Polaris Missile program. Through its use, the Polaris program was completed 18–24 months ahead of the originally scheduled completion date. Since 1959 Pert has been successfully used in almost every type of large-scale industry. For example, Pert analyses have been used in a number of areas of the computer industry, in the cinema-film production industry and particularly in the US military. In fact, a Pert analysis is required by the US military for all large contracts. A twin to Pert is Gert (graphical evaluation and review technique) and the reader is referred to Lockyer (1977) to obtain a reasonable comparison of the two.

The critical path method (CPM) is another planning and coordinating tool and is very similar to the one described in the text. It was developed in the construction industry where previous experience was used to obtain time and cost estimates of the various phases of the project. Its development was jointly sponsored by E.I. du Pont de Nemours & Co. and the Sperry-Rand Corporation. It was applied first to the construction of a chemical plant and then to the shutdown of the plant for overhaul and maintenance.

All of these are useful for organizing the activity sequences necessary for the completion of a large-scale design project. These methods are adequate to define and control a rigid sequence of well-defined activities but often lack the flexibility needed for planning in complex large-scale systems. This is because the conventional activity flowcharts do not allow for alternatives, decision points and logic. The type of charts which do allow for these requirements include delta charts and Gantt charts. Gantt was a pioneer in industrial engineering of time and motion study, and the Gantt chart, by virtue of its organization, allows the manager to tell at a glance the time at which certain phases of the project will be completed, which are in progress and which are shortly to commence. They are also helpful to all concerned with the project, even those who are only involved on a part-time basis, as a consequence of its elementary design. There does not, however, appear to be a convenient way to depict options and uncertainties on a Gantt chart as there is with the delta chart. The latter takes its name from the acronym formed from the initial letters of five graphic elements used in constructing the chart: decision box, event box, logic box, time arrow and activity box. The other graphic elements are used in addition to these, namely the connection box (to depict a large fan-out of activities) and the ground termination (an earth symbol denoting the termination of a sequence of preceding elements).

A full discussion of these charts is inappropriate within this text, but the reader is referred to Lockyer (1977). This also contains an excellent treatment of the methodology for large-scale systems extolling the rationale behind systems engineering as befits its unique application to the solution of the complex, large-scale problems which society faces today. As it does so, explaining the close interrelationships between technology and policy, it rather appropriately discretely emphasizes the virtues of network analysis as an integral part of the systems engineering approach, underlining the thoughts put forward in this chapter.

7.7 EXERCISES

1. State Bellman's principle of optimality and explain very briefly how it may form the basis of optimal decision-making. The current through a series RL circuit is required to be increased from zero to 1 A in $2T$ seconds. To achieve this, a voltage of magnitude $V(0)$ is maintained across the circuit for T seconds immediately after which the magnitude is changed to $V(T)$ and maintained for a further T seconds.

The transient response of this current is represented by the iterative equation:

$$i\{(k+1)T\} = 0.9i(kT) + V(kT)$$

Calculate the optimal value of $V(0)$ and $V(T)$ to minimize the given cost functional:

$$J_2\{i(0)\} = \min_{V(0)} \Sigma_{k=0,1} \{i(kt)^2 + V(kT)^2\}T$$

2. A manufacturer transports his goods to various depots by sea, road, rail and air. For a given distance the cost is proportional to the square of the load carried, where a, b, c and d are the respective constants (£/tonne^2) for each mode of transport. Calculate the allocation of a load of 1,000 units which will minimize the cost of a particular shipment. What is the minimum cost a manufacturer can expect to incur for shipping this load when $a = 2b = 3c = 4d =$ £10/tonne^2, if he uses an optimal policy?

3. A certain random experiment has four outcomes X1, X2, X3 and X4. Calculate the probability of each outcome if the probability of the sequence of independent events X1, X2, X3, X4 over four successive trials of the experiment is to be a maximum. What is this maximum probability?

4. State the principle of optimality and relate this to the optimization of multi-stage decision processes. Using this principle, show that a necessary and sufficient condition for the total power loss in Figure 7.12 to be a minimum is that $V_1 = V_2 = V_3$.

5. A three-stage rocket firing system, with parallel redundancy and probability of successful operation for the assemblies in each subsystem, is shown in Figure 7.13. Show that the reliability of the system, r, is given by:

$$r = p_1p_2p_3\,(2 - p_1)(3 - 3p_3 + p_3^2)$$

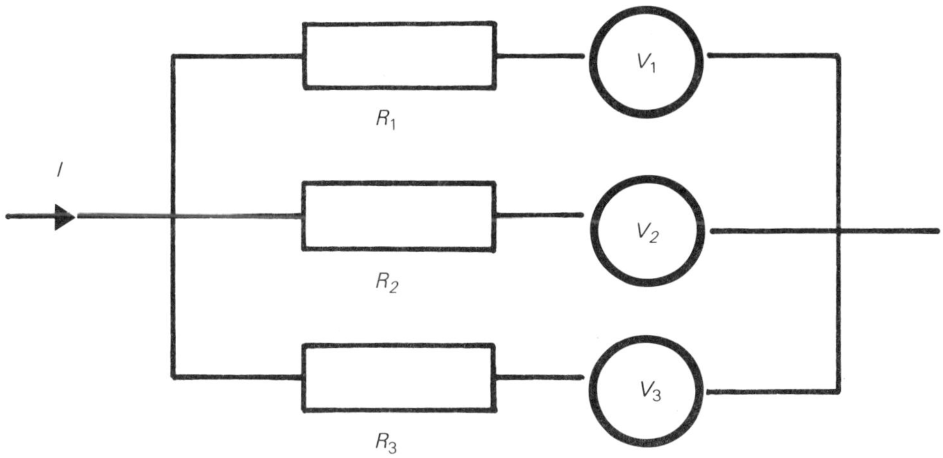

Figure 7.12

Assume now that $p_1 = p_2 = p_3 = 0.5$ and that the cost of assemblies are 1, 2 and 3 units each for stages 1, 2 and 3 respectively. If a capital sum of 9 units is available, determine, by the method of dynamic programming, the number of assemblies for each subsystem which will maximize the reliability of the total system.

6. Compare the methods of linear and dynamic programming in terms of optimization problems generally associated with large-scale systems design. State to which of these methods the following equation belongs:

$$f_i(W) = \max_{0<n<[W/w_i]} \{(nV_i + f_{i-1}(W - nw_i))\}; \qquad (i > 1)$$

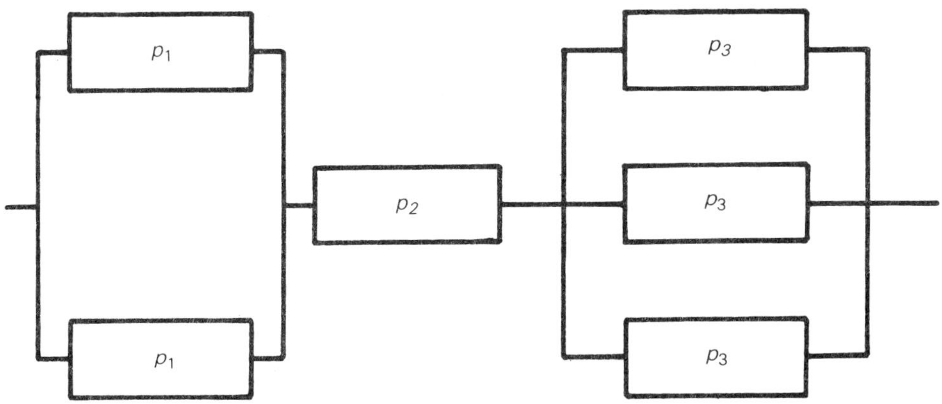

Figure 7.13

A limit of £100,000 has been set on supplying three types of cable to feed a particular substation. The rating and cost of these cables is shown in Table 7.10. Using the principle of optimality, determine the number of each type of cable so as to provide maximum power transfer capability to the substation.

7. The network model shown in Figure 7.14 represents a motorway system which a traveler proposes to use in moving from point 1 to point 11. Calculate, using the method of dynamic programming, the particular motorway sections which the traveler must take in order for his journey to be of minimum distance.

8. An electricity supply authority decides to build an overhead line between points A and B, Figure 7.15. A number of wayleaves (rights-of-way) defining intermediary points suitable for the location of substations are suggested. The distances for each wayleave are marked on the network. What route will the overhead line take if a minimum distance criterion is applied? Use dynamic programming to calculate this.

9. For the network shown in Figure 7.15 calculate the minimum distance route, using Minty's algorithm.

10. For the network shown in Figure 7.15, calculate the maximum flow and the final flows in the network using the Ford–Fulkerson algorithm. (Assume that 'distances' are 'capacities' and that flows relate to oil volumes for a pipeline network.)

11. Confirm that the minimum distance route of Figure 7.2 is 'ACEILN' using Minty's algorithm.

12. For this same network calculate the maximum flow and the final flows using the Ford–Fulkerson algorithm, assuming that 'distances' become 'capacities' and that 'flow' relates to traffic volumes.

13. With reference to the network of Figure 7.15 use the matrix method to calculate the minimum distance route between nodes 2 and 10, and nodes 1 and 11.

Table 7.10

Cable type	Rating (MV A)	Cost (£10^3)
A	38	29
B	27	22
C	19	16

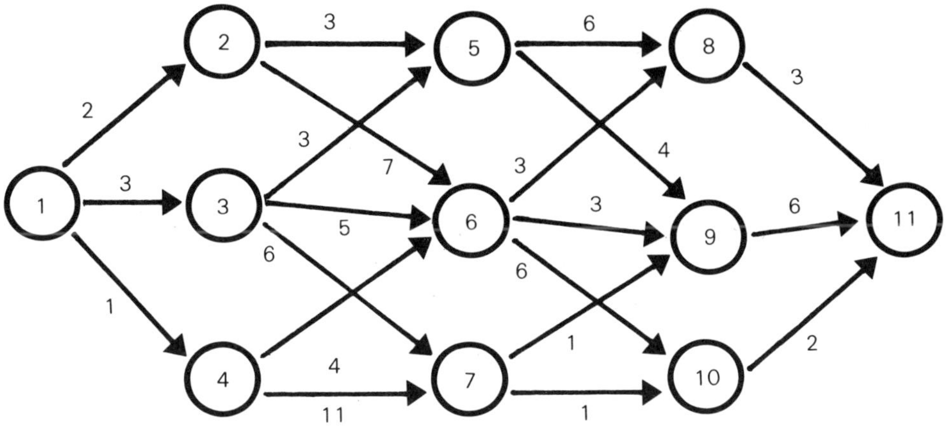

Figure 7.14

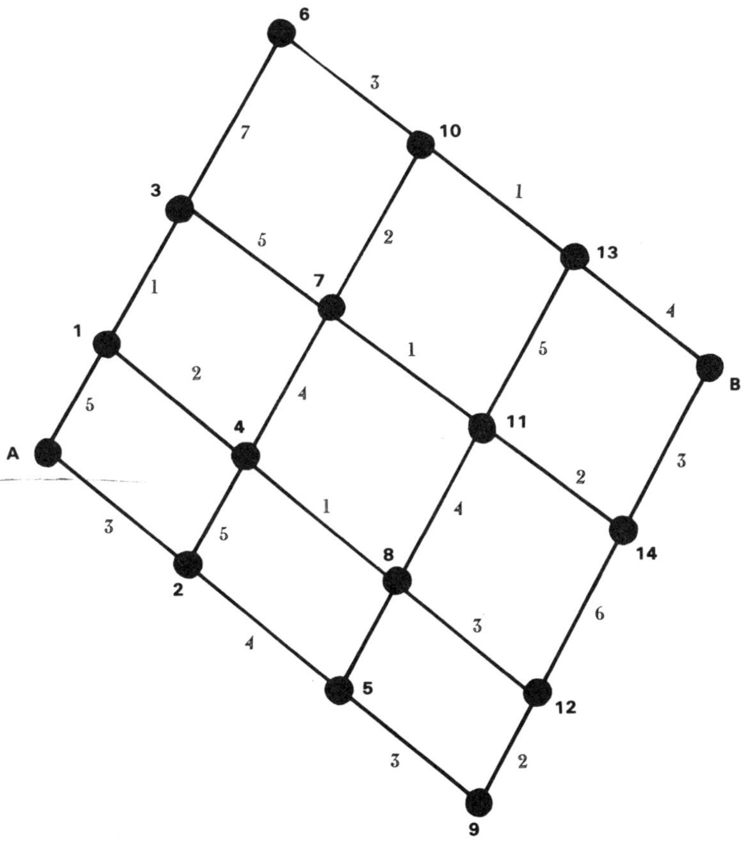

Figure 7.15

14. Solve the assignment problem of Table 7.11 using the branch-and-bound algorithm described in Section 7.5.2.

15. Use Eastman's branch-and-bound algorithm. described in Section 7.5.2, to solve the traveling salesman problem with a distance matrix as shown in Table 7.12.

16. Assume the figures given against the activities in Figure 7.9 to be capacities and the network itself to be not a project model but a supply water network. Using Minty's algorithm, calculate the maximum capacity route. Using the labeling algorithm, calculate the maximal flow into node 5. (Ignore the dummy activity in both calculations.)

17. A project contains seven jobs represented by Figure 7.16. The duration of each job, in hours, is indicated. Determine the critical path.

The planning department finds that an extra job has to be inserted between events 2 and 5. Of what duration will this job need to be, if it is to become a critical activity?

18. An engineering project can be represented by Figure 7.17. Before the project commences, however, it is found necessary to include an extra job, duration four days, which depends only on the completion of activity (2, 8). Activity (3, 4) will depend on this new job, otherwise the project remains unchanged. What will the revised network look like, and what is its critical path?

Table 7.11

Operatives	Task 1	Task 2	Task 3	Task 4
A	2	10	9	7
B	15	4	14	8
C	13	14	16	11
D	4	15	13	9

Table 7.12

From city	To city 1	2	3	4	5
1	∞	8	5	8	6
2	1	∞	3	1	8
3	1	7	∞	6	7
4	3	4	6	∞	9
5	6	5	3	7	∞

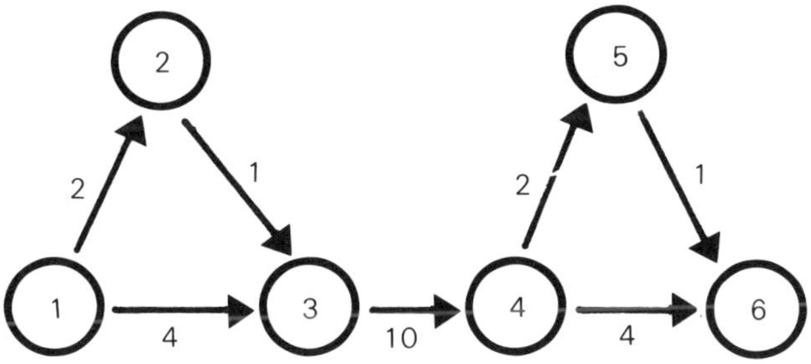

Figure 7.16

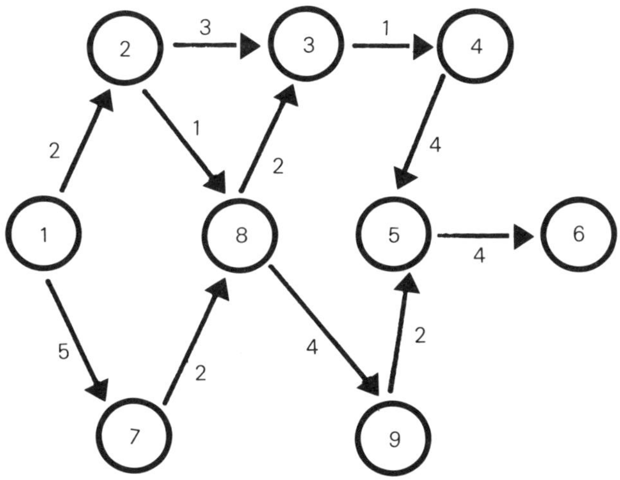

Figure 7.17

7.8 FURTHER READING

Bellman, R. (1957) *Dynamic Programming*, Princeton University Press.

Boardman, J.T. 'Design of dynamic programming feedback controllers for multivariable time-invariant linear systems', *Int. J. Control* **17** (1), 33–43.

Boardman, J.T., Garrett, C. and Robson, G.C.A. (1986) 'A recursive algorithm for the optimal solution of a complex resource allocation problem using a dynamic programming formulation', *The Computer Journal*, **29** (2), pp. 101–9.

Carre, B. (1980) *Graphs and Networks*, Clarendon Press.

Cooper, L. and Cooper, M.W. (1981) *Introduction to Dynamic Programming*, Pergamon.

Dreyfus, S.E. (1977) *The Art and Theory of Dynamic Programming*, Academic Press.
Helman, P. and Veroff, R. (1986) *Intermediate Problem Solving and Data Structures*, Benjamin/Cummings.
Kaplan, E.L. (1982) *Mathematical Programming and Games*, Wiley.
Lockyer, K.G. (1977) *An Introduction to Critical Path Analysis*, Pitman.
Norman, J.M. (1975) *Elementary Dynamic Programming*, Edward Arnold.
Smith, D.K. (1982) *Network Optimisation Practice, a Computational Guide*, Ellis Horwood.
Wilson, R.J. (1985) *Introduction to Graph Theory*, Longman.

8 Systems engineering and expert systems

8.1 AIM AND OBJECTIVES

This final chapter has three objectives. The first is to provide a description of expert systems. These are relatively new developments in computer software which form part of artificial intelligence (AI). In recent years some exciting results have emerged from AI which are now beginning to be pulled through into industry and commerce, whose more enlightened members have realized the tremendous potential benefits that can be obtained from such developments as expert systems (Feigenbaum *et al.* 1988).

A second objective is to present the results of a practical piece of research in which the author and his research colleagues have been actively involved for the past five years (Barber 1988, Marshall 1988). In this research, expert systems have been developed in close collaboration with industry who foresaw these systems as essential components in their future line of business, which rested primarily on a systems engineering approach. Two examples of this collaboration are cited: first, the manufacture of flight simulators; and, secondly, the creation of bid proposal documents for the process of tendering for process plant construction.

A third and final objective is to examine the interrelationships which already exist, and those which are likely to be developed as the field of expert systems grows, between it and systems engineering. Conceptually at least, each has much to gain from the developments and extensive application of the other. As the complexity of the larger-scale systems with which the system engineer must deal increases, it is highly likely that these knowledge-processing tools, many having been forged in the cauldron of engineering applications (Waterman 1986), will become the stock-in-trade for the future systems engineer.

The overall aim of the chapter is to reinforce the concepts which are common to both expert systems and systems engineering. By relating this mutual interaction it is hoped to fire the imagination of practicing systems engineers, be these apprentices or masters in their profession, by pointing to the many exciting opportunities that lie ahead as a result of new areas being opened up by subjects such as AI.

8.2 WHAT ARE EXPERT SYSTEMS?

An explanation of what expert systems are and what they can do is best advanced by first giving a general overview of the subject of AI. If we examine some of the goals that the proponents of AI have set themselves we can get a clearer picture of the mission for expert systems. Pat Winston (1977) in his book *Artificial Intelligence* has articulated the goals of AI with great clarity and poignancy. Winston writes (with italics by the present author):

> Artificial Intelligence (A.I.) is the study of ideas which enable computers to do things that make people seem intelligent. The central goals of A.I. are to make computers more useful to *understand the principles* which make intelligence possible.
>
> A.I. enthusiasts believe that using computers to *understand the central issues* and the dimensions of intelligence is a powerful addition to the traditional methods of psychology, philosophy and linguistics.
>
> A.I. excites people who want to *uncover the principles* that apply to all intelligent information processors, not just those that happen to be made of wet nervous tissue instead of dry electronic gadgetry. Consequently there is neither an obsession with mimicking human intelligence, nor a prejudice against using methods that seem involved in human intelligence.
>
> The overall result of this new point of view may be new ideas about how to help people become more intelligent. The methodology involved in

> making smart computer programs may transfer to making smart people smarter.

The point of the italics is to highlight an affinity for Winston's philosophy for AI, namely that it is a major development in the history of computing which deserves and demands the attention of those who have a professional responsibility to *add* to their stock of knowledge. The reason is simple: AI promises to make this augmentation process more efficient and thereby enable the professionals to serve mankind more effectively. Engineers, quite clearly according to the philosophy expounded in this book, are marked out for singular attention amongst this body of professionals.

We continue our commentary on AI by examining a particularly exciting development in computer software, and one that will be of chief interest to systems engineers, namely expert systems, sometimes referred to as intelligent knowledge-based systems (IKBS). The overriding feature with these systems is their capacity to capture and exploit knowledge, thereby endowing them with an essential cognitive ability.

We begin our description of expert systems with an account of the seminal work in this area carried out at Stanford University. In the mid 1970s, the Mycin project received great acclaim. The researchers, Shortliffe and others (Buchanan and Shortliffe 1984) had been working on attempts to make computer programs tackle problems hitherto untried, and which were going to require human knowledge to be encoded within them to work at all. The problem posed for the Mycin system was to diagnose microbiological infectious diseases and then to prescribe a treatment in terms of drugs and their dosages. The system was to behave like the physician; after a consultation to establish a particular patient's data, the system had to use the physician's knowledge and hence arrive at acceptable conclusions.

The designers took a major step in trying to separate the knowledge from the inferencing procedures, thereby creating what is considered today to be the classic expert system's architecture of knowledge base and inference engine (or mechanism). Their belief was that the inferencing procedures should contain no knowledge and therefore be domain-independent. All the knowledge was made explicit in the knowledge base alone. If that were true then it would be possible to remove the knowledge base and incorporate another, from a different domain, thereby creating a new expert system. This they did. The knowledge from Mycin was emptied, and the result, which was referred to as a shell, became known as Emycin (essential Mycin). The team recoded Mycin in Emycin to prove it would work, and it did. Since that time, much more work has been done by many prominent AI researchers, and the community has seen an escalation of the interest and success of such systems. Today, there are a number of commercially available shells and AI languages, which will run on machines ranging from mainframes to micros. There is a wide range in terms of price, complexity and capability amongst these systems.

All expert systems have at their heart a means of representing information which can adequately describe the state of their world. This information effects a mental model of the world, and the systems have the ability to process the model in order to make decisions about the world. The model of reality is expressed using a language of internal representation; the complexity of which can vary enormously depending upon how it is intended to be used.

Expert systems which have sensory input translate the information they receive into the internal representation, for it is in this form that it can be centrally processed. Until recently it has been uncommon for expert systems to possess a sensory input, but with the increasing robustness of blackboard architectures (Penny Nii 1981) expert systems are beginning to appear which have real-time planning and control capabilities (Raj Reddy *et al.* 1982, Hayes-Roth 1983). A more common transfer of information between the expert system and its environment is by question and answer sessions.

Since the internal representation language is used to build the model of the world for processing, its power of expression is of major importance. Imagine for example a robot observing two scenes, one of a car parked at the roadside, the other of a car hurtling towards it. If the internal representation was only able to express 'a car', it could be fatal for the robot! Clearly the internal representation must never be at a loss to express any crucial relationship or fact. A poor internal representation might prevent certain information from being expressed which could seriously impair performance. Whatever the choice of internal representation, ambiguity cannot be tolerated as it would lead to false results (Charniak and McDermott 1985).

The state of the world is expressed using the internal representation, usually as a collection of facts. The collection is termed the database and is simply the place where all such information is held. The operation of an AI system will repeatedly require the database to be searched to find relevant information for processing; which raises the issue of speed efficiency. Any delays incurred while searching the database will have a significant impact on the system's ability to deliver results quickly.

The choice of internal representation can have a considerable impact on the efficiency of the search. Whilst some representations might be better suited to modeling a domain than others, they could be very expensive to search and process. For instance, English could be used as an internal representation which with its richness, might describe a domain very eloquently. However, its complexity and loose definition would mean that efficient processing would be almost impossible. There is therefore a trade-off between the ability of the internal representation to express the content of the domain,and the system's ability to search and process it.

We have now encountered the key issues of AI systems: internal representation – the language used to describe the state of the domain; the database – to store all of the information; search – to find the relevant information; and processing – to achieve the desired results. In the context of expert systems development, it can be convenient to make a clear distinction between the terms 'information' and 'knowledge'. This may at first seem frivolous, but there is an important differentiation. Information is that which is required to allow decisions to be made, whereas knowledge makes use of information (as evidence in a sense) in order to make decisions or draw conclusions, the latter constituting additional information. A simple analogy is with a control system in which the information is the statement of the world conditions, and the knowledge is the controlling element making decisions to bring about a desirable state of world conditions. This philosophy has been fully reflected in the methodology outlined in this author's research. The information system is represented graphically by an associative network and textually by its equivalent form, the frame-base. The controlling element or decision-making capability is provided by a rule-base. The details of the architecture, philosophy and methodology are contained in ensuing sections. Before these are presented however, it may be worthwhile rehearsing a few elementary principles of knowledge and information representation. First it is pointed out that rule-based production systems are the dominating technology in expert systems. Rules allow knowledge to be explicitly stated in a form which is easily understood, and readily processed by the system. More sophisticated rule-based systems facilitate reasoning with uncertain data and incomplete knowledge, and are able to explain their line of reasoning and justify their conclusions. Generally speaking, the collection of rules constitutes a knowledge base, and an 'inference engine' will employ specific rules under qualified conditions in order to arrive at a desired result.

A rule is a list of premises followed by a conclusion. A premise is a test which is satisfied by recourse to the information base. A conclusion is a means of adding to the information base. Note that already a degree of circularity is possible. In attempting to execute a rule, i.e. reach a conclusion, its premises must be satisfied and this is conditioned by the state of the information base. It may prove necessary to add to the stock of information, by depositing into it conclusions reached by other rules. One is hopeful that the circularity is neither paradoxical nor vicious! A rule therefore is a means of supplying information in specific cases. The inference engine is a procedure which takes a list of rules as an argument. When it is given a particular argument, it attempts to exercise the rules which it has been given, and if it is successful then it yields the desired result (which the 'originator' believed could be obtained by the application of those rules).

Now it just seems to be an inescapable fact of life that rules are necessary. I am a person with creature comforts and I know when these are not being met, and I know what to do to get them met. My fundamental rules of life might be: 'avoid danger', 'keep healthy', 'have fun' and 'help people'. These will be operating at various times and in marvellous ways throughout my entire life, and they perform best when prompted by circumstance. What prompts the inference engine? Initially it is a request for information, which is why the bottom line of all explanation is: 'because you asked me!' Therefore it should come as no surprise to learn that techniques for finding out information are the essence of expert systems methodology. Furthermore, no one should show alarm when told that an expert system, just like an expert, sometimes needs to ask in order to know, and must be permitted to say 'I don't know' when that is the case.

From an implementation viewpoint, an inquiry procedure becomes the prime mover of calls on the information base, which amount to the solicitation of values from a highly structured database. When these values are not directly available from established or default information, and cannot be inherited from class hierarchies, then they can be supplied by simple computer programs (procedures) specifically allocated to points in the database. This is known as the method of procedural attachment. One procedure is to 'ask', another is to 'calculate', but yet another, and a feature unique to this methodology, is to 'infer'. In this situation the inference engine, a procedure like any other, is attached to the database as the designer directs, thus providing the information base with an inherent capability of self-referential treatment as directed by rule-sets. Thus the distinction between knowledge and information is contrasted by this unusual integration of the two.

The conclusion which a rule may make is not confined to simply being a value. It may in fact be an object of significance in the world model (generally referred to as a concept). In this case the rule is making a link between two concepts. The system described in the following section is capable of this and approaches it from three standpoints. First the concept or object may already be in existence, in which case all that is necessary is simple connection. The object to which connection must be made is selected from the candidate set whose scope is set with a special function termed 'each'. A second case is when the object does not exist but must be instantiated from its generic object. It then remains only to determine whether the instantiation is single or multiple, for which the special functions 'a' and 'some' were invented. A final case is where there can be some doubt whether the instantiation required has already been made. For example 'Fred lives in a house, Fred is married to Mary, and Mary lives in a house: Does Fred live with Mary?' The system resolves this dilemma either by 'instantiation directives' (lists of information stored in a reserved location belonging to the object) or by 'manual naming', i.e. the questioner or user informs the system of the particular instantiation to be made. Further details of all these features are contained in the following section which gives a detailed description of an original expert system development environment.

8.3 THE RULE-BASED FRAME SYSTEM

The rule-based frame system (RBFS) uses an internal representation which is well suited to describe many domains and which may be searched and processed efficiently. The novel architecture of the RBFS integrates three well-known representation schemes, namely frames, associative networks and rule-based systems.

Historically the initial experimentation was with the frame scheme alone and attention focused upon the problems associated with using it to build an expert system. It was apparent that building domain models and processing them effectively would require improvements if the frame system were to be readily employed. To this end an investigation was made into two other well-founded areas of AI research, associative networks and rule-based systems. All of these techniques have been used in many systems, but they had never before been truly integrated, although in some systems they do co-exist. For increasingly tougher requirements the advocates of any particular technique always seem to build on that technique which they favor most (naturally enough). However, it could be that the nature of the problem of building expert systems lends itself more to a variety of representation techniques. No one wants to build a system which is an ad hoc collection of techniques, but if a philosophy can be formed directing the use of each technique within the scheme, and this can be reflected in the architecture, then the result might be more than could be achieved through the extension of any individual technique alone. The result of this work was RBFS, a system in which the integration of these three techniques seems, in retrospect, surprisingly natural.

8.3.1 Frames

A simple form of internal representation is the 'object–attribute–value' (OAV) triple in which all the information modeled in a domain is expressed in the database as a list of simple triples:

```
[  [Harry       loves               Mary]
   [house_2     location            Brighton]
   [Harry       is_married          true]
   [Harry       owns                house 2]
   [house_2     no_of bedrooms      3]
   [Harry       hobbies             jogging]
   [Harry       hobbies             swimming]  ]
```

To use this database two very simple fundamental procedures are needed: one to retrieve facts called 'inquire' and the other to add new facts called 'put'. The former is of more interest since it is responsible for the database search. It takes arguments for the object and the attribute, and returns the desired value; for example: inquire ("Harry", "loves") yields "Mary". Normally, there would be a large collection of such facts in a database in order to describe the state of the domain completely. In such a situation, the simple list structure of the database incurs a search overhead, since each inquiry requires a scan through the database looking for a match on the object and attribute fields. The search time of this scheme is proportional to the number of facts stored. A more sophisticated approach is to collect together all the facts associated with each object, and bind all the attribute-values of that object into one unit called a frame. Then the database becomes a collection of frames:

```
[Harry     [loves               [Mary]]
           [owns                [house2]]
           [is_married          [true]]
           [hobbies             [jogging swimming]]]

[house2    [location            [Brighton]]
           [no_of_bedrooms      [3]]]
```

This formation was founded by Minsky (1975) and was termed the frame-slot notation, in which the frame represents the object (or concept) and the attributes are the slots. The scheme partitions the database into concepts; consequently the search procedure 'inquire' can go directly to the frame, and only has to search for the particular slot – the search time is then nominally proportional to the number of slots in the frame.

However, although this scheme gives an ability to index into the frame, it does not have the ability to index into the value. Consequently questions like 'who loves Mary?' are much more difficult to answer than questions like 'whom does Harry love?' In the former case all frames would have to be searched for a slot called 'loves', and those found checked for the value 'Mary'. Essentially the scheme lacks a 'back-pointer'; facts are only recorded one way. The problem is solved by the addition of the 'inverse' relation. Whenever a fact is added to a frame, if the value is itself a frame (as in the case of Mary), then the inverse relation to the value-frame must be added, thus:

```
put ("Harry", "loves", "Mary")
put ("Mary", "i_loves", "Harry").
```

The 'i_' prefix indicates the inverse relation. In the 'Mary' frame it indicates that she is loved by 'Harry' but says nothing of her love for him. Now the previous difficult questions are asked easily, using the same 'inquire' procedure, e.g. inquire ("Mary","i_loves"). Below, the example frame-base is modified to account for the inverse relations. These are only added when the value of a frame-slot is itself a frame, simple values like 'true' and numbers, are not frames and therefore do not have the inverse link.

```
[Mary       [i_loves          [Harry]]]

[Harry      [loves            [Mary]]
            [owns             [house 2]]
            [is_married       [true]]
            [hobbies          [jogging swimming]]]

[house      [location         [Brighton]]
            [no_of_bedrooms   [3]]
            [i_owns           [Harry]]]

[Brighton   [i_location       [house2]]]
```

There is scope for a further improvement of the internal representation. Consider two pieces of information that we might like to store:

1. We have been told that Harry loves Mary.
2. A reasonable assumption is that Harry is right-handed.

There is a difference here between the type of information we want to represent. There are those facts which we know unequivocally, and those we can make a reasonable guess about. In human reasoning we make extensive use of such default information, it gives us an ability to consider a problem with only sketchy information. Ideally therefore, an IKBS should be equipped with the same capability, but it must be aware of the type of information it is using.

This is accomplished by giving the slots attributes, called facets (Winograd 1975). The 'value' facet means the fact is known; and the 'default' facet means that it is a reasonable assumption:

```
[Harry    [loves        [value     [Mary]]]
          [owns         [value     [house2]]]
          [is_married   [value     [true]]]
          [hobbies      [value     [jogging swimming]]]
          [dexterity    [default   [right_handed]]]]
```

When the knowledge engineer builds the model of the domain, he is able to state all the default information, thereby enabling the user to specify only the exception when it occurs. This minimizes his input and provides the system with a common-sense information of the domain. Obviously if the exception data do arrive then they must be stored with the 'value' facet, e.g. if the system were told that 'Harry' was really left-handed, that part of his frame would appear:

```
[Harry [dexterity [default    [right_handed]]
                              [value [left_handed]]]]
```

The 'inquire' procedure must be redefined to look for the default facet in a frame's slot only in the absence of a value facet. Now suppose the system were told that 'Harry' was right_handed, the frame would change as shown below, and subsequent inquiries of Harry's dexterity would yield [right_handed left_handed] – ambidextrous!

```
[Harry [dexterity [default    [right_handed]]
                              [value [left_handed right-handed]]]]
```

Consider two more facts about Harry; he earns a salary of £10,000, and has a tax code of 352. For the system to process the inquiry, "how much tax does Harry pay?", the slot 'tax_payable' would have to hold the calculated sum. But suppose that Harry receives a wage award, then the 'tax_payable' value must be recalculated and the slot updated – which is rather clumsy. A better approach is not to store the calculated value, but to store the name of a procedure (e.g. 'calc_tax_payable') which when run will calculate the value. The system needs to know that 'calc_tax_payable' is in fact the name of a procedure, not the value, and this is facilitated with a facet called 'if_needed':

```
[Harry                        [salary [value [10000]]]
                              [tax_code [value [352]]]
                              [tax_payable [if_needed
                                [calc_tax_payable]]]]

define calc_tax_payable();
                              = 0.29*(inquire ("harry","salary")
                              −10*inquire ("harry", "tax_code"))
end procedure;
```

The facility for 'procedural attachment' is a powerful feature. It allows a database to contain essential values from which other values are calculated automatically. Accordingly the procedure 'inquire' is modified so that in the absence of 'value' and 'default' information, if an 'if_needed' entry is present, the procedure is executed and its result returned. An extension in RBFS allows procedures to take arguments, enabling individual procedures to be more flexible and widely used. In the example

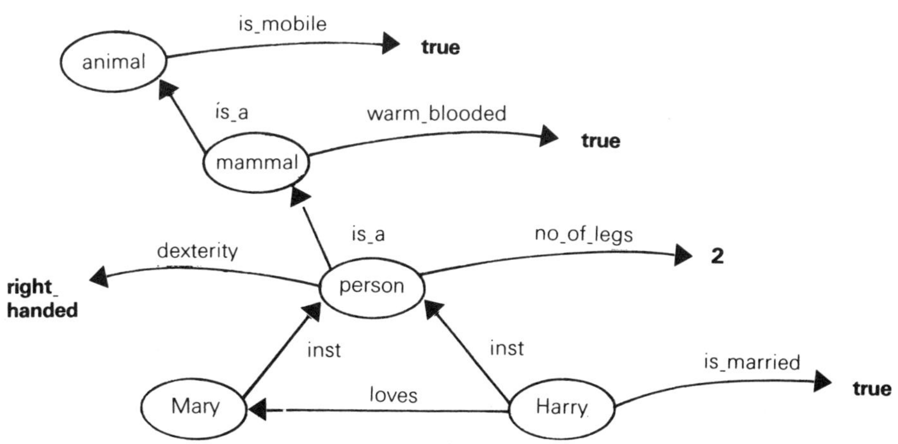

Figure 8.1 A simple associative network.

presented thus far, there are two characters: 'Harry' and 'Mary'. Yet much of what could be modeled for each is the same, e.g. by default they have two legs and are right-handed, because they are both humans. This feature is exploited by modeling the class as well as the individual. The idea is extended by the development of class (or type) hierarchies: 'Harry' and 'Mary' are instances of the concept 'person', which is a 'mammal', and 'mammal' is an 'animal' and so on; Figure 8.1.

The class hierarchy is modeled by two special relations, 'inst' and 'is_a', which have mathematical meanings of element and subset respectively. The classification scheme is not only conservative on memory, but a natural way of describing the world. Again, the inquire procedure is redefined to exploit the classification structure, so that information which is not explicitly stated in an individual's frame can be inherited from the class hierarchy to which it belongs.

```
inquire ("Mary", "dexterity") =
look for Mary's dexterity – value, default, if_needed
if nothing found
look for class_of (Mary) dexterity – value, default, if_needed
if nothing found        look for class_of (class_of (Mary) ...
```

The preceding discussion has shown the evolution of the structure of the database and the advantages that structure brings. These features alone constitute a powerful and flexible system of representing and retrieving information within a domain, and are characteristic of frame systems. For examples of sophisticated frame systems which have been developed see Bobrow (1975) and Goldstein and Roberts (1984). Details of how to implement such a frame system are given in Winston and Horn (1980). RBFS has been implemented with all the features described and includes a comprehensive set of retrieval and manipulation procedures.

8.3.2 Associative networks

The associative networks model (also known as semantic networks) is a scheme in which the facts about the domain are expressed as a directed graph with nodes and links which respectively represent concepts, and the relationships between them. They are popular with many AI researchers and a useful reference is Findler (1979). In terms of implementation, the graphical networks are typically held internally as a collection of data structures allowing indirect associations. But clearly the network could easily be expressed using the frame scheme, where the frames are the nodes and the slots are the links. The two representation

```
[animal      [is_a [value [mammal]]]
             [is_mobile [value [true]]]]

[mammal      [i_is_a [value [man]]]
             [is_a [value [animal]]]
             [warm_blooded [value [true]]]]

[man         [i_inst [value [Harry] [Mary]]]
             [is_a [value [mammal]]]
             [no_of_legs [default [2]]]
             [dexterity [default [right–handed]]]]

[Harry       [inst [value [person]]]
             [loves [value [Mary]]]
             [is_married [value [true]]]]

[Mary        [inst [value [person]]]
             [i_loves [value [Harry]]]]
```

Figure 8.2 An example frame-base.

schemes are equivalent, but the frame scheme has the additional features of procedural attachment and inheritance. Perhaps the greatest virtue of associative networks, but one which seems to be little exploited, is their ability to be naturally depicted as a graph, giving an invaluable visual portrayal of the database. Compare the frame example of Figure 8.2 with the equivalent associative network in Figure 8.1. Which is more comprehensible?

If humans relate more readily to a visual representation then why not allow information networks to be developed graphically? The idea was appealing. Consequently, RBFS is equipped with an interactive tool to develop networks graphically. This is known as VEGAN (Visual Editor for the Generation of Associative Networks), and for details of this the reader should consult Kellett (1986). VEGAN provides a number of powerful facilities for the development and maintenance of large networks, enabling the user to sketch the domain model at the terminal. A mouse is used to position the network, which is built up with simple operations like add/delete node or link, and correspondingly the frame-database is maintained. Not surprisingly, the addition of this tool brought about a significant improvement in the ease-of-use of RBFS. However, as might be expected when encoding large complex models, the visual display becomes crowded and increasingly difficult to comprehend. Even though the tool offers zoom and pan facilities, large networks tend to become unmanageable. As a part of the research effort Vegan was extended with the implementation of a number of interesting ideas:

1. Most of the time the user does not want to see the whole network, since it presents more information than he can cope with. Instead he may wish to restrict his vision to just a range of node and link types, which may be displayed in different colors. This has been termed 'information filtering'.
2. A number of filters give different views of the same network, and if these can be arranged as planes, the user can quickly move between planes.

These ideas have allowed Vegan to manage networks an order of magnitude more complex than was previously possible, i.e. over 400 nodes and 800 links. The visual aspect should not be underestimated, for it is of considerable use to the knowledge engineer and expert when trying to develop large domain models. Indeed it is doubtful whether some of the large models could have been maintained by any other means. Vegan has certainly proved itself to be a valuable knowledge engineering tool since it eases the process of encoding the information within the domain.

Now that information networks can be seen, it becomes clear that the 'inquire' facility previously offered should be extended to deal with questions like 'who is Harry's brother's spouse?' and 'who does Harry's love, love?' These are composite questions. The second could be answered by:

```
inquire ("Harry","loves") → x:      (where x would be Mary)
inquire (x,"loves") → y:            (where y is whom Mary loves)
```

Inspection of these kinds of questions reveals that they are merely traversals of the information network over a number of links (the present inquire facility traverses just one link – the slot in question). To handle these questions, a new inquiry procedure was defined, 'c_inquire', which takes as an argument a 'path'. The path is defined from a set of start nodes followed by a list of links to traverse. Thus to answer the questions above one would write the following two statements:

```
c_inquire ([[[Harry] brother] wife] →
c_inquire ([[[Harry] loves] loves]) →
```

The c_inquire procedure facilitates an extremely powerful search capability which needs to be appreciated. To illustrate the example the domain is developed as shown in Figure 8.3 and the following shows how

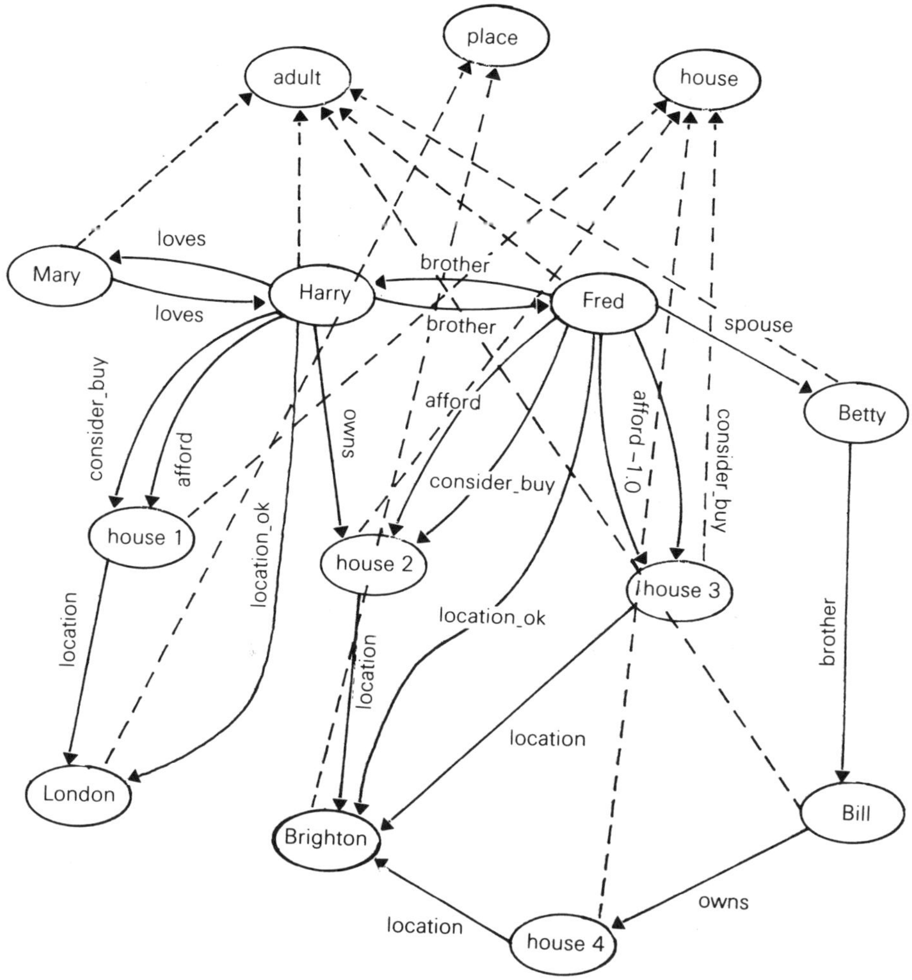

Figure 8.3 The house buying example.

a number of questions would be phrased. The inquiry 'what does this person's brother own?' is written:

```
c_inquire ( [[[this-person] brother] owns])
```

When 'this-person' is equated to Fred the traversal yields house2, and when it is Betty the result is house4. Similar inquiries such as 'what is the cost of the items which this person's brother owns?' can easily be formed and evaluated thus:

```
c_inquire ([[[[this-person] brother] owns] cost])
```

Inquiries can return multiple values because although the argument to inquire is a path, the search may take the form of a tree, e.g. 'who are all the house owners?':

```
c_inquire ( [[[house] i_inst] i_owns])
yields [Harry Dick]
```

The 'c_inquire' procedure is evaluated by successive application of the 'inquire' procedure, which therefore incorporates the facet and inheritance capabilities. An example of how this is exploited is, 'how much tax does the person considering buying Harry's house pay?':

```
c_inquire([[[[Harry] owns] i_consider_buy] tax_payable])
```

The first three terms yield 'Fred', whereupon the system looks for the 'tax_payable' slot via a 'value', 'default' or 'if_needed' facet, but none exists. Thus, 'inquire' would look up the class hierarchy to ask the question at the 'adult' node, where it executes the procedure 'calc_tax_payable' to get the result. Inheritance is transparent and could occur anywhere on the path, and even several times on a path.

Clearly this is an extremely effective way of expressing a general inquiry from the model. The tree search of the frame-database readily facilitates very complex information retrievals; increasing the power and capability of the two technologies beyond that which either could have achieved alone. Furthermore it should be no surprise that there is a high degree of correlation between the English questions and the path description, making it easy to phrase inquiries and implying that at some future date, a natural language front end could be added.

8.3.3 Production rules

In the increasing number of expert systems being developed, the dominating technology appears to be that of rule-based systems. Rules allow knowledge to be explicitly stated in a form which is easily understood, and readily processed by the system. More sophisticated rule-based systems possess the following:

1. Reasoning with uncertain data and incomplete knowledge.
2. Justification of their conclusions.
3. Explanation of their line of reasoning.

Not surprisingly, such useful capabilities have made them a favorite paradigm with many researchers. Indeed, they were considered an essential requirement to facilitate the processing of information held in a frame-database. But how could rules and frames be integrated? A crude solution would be to design two, essentially autonomous, subsystems of frames and rule-based technology, and to establish a means of communication between the two. However, in the course of the author's research, a much more fundamental means of integration was found. In order to present this, it will be necessary first to examine the classical architecture of a rule-based system, which essentially comprises three elements:

1. *The knowledge base*: a collection of rules which have been elicited from the expert.
2. *The database*: all the facts that are known defining the current state of the problem and facts regarding the domain in general.
3. *The inference engine*: given an objective to find out something, it will search for relevant rules in the knowledge base and apply them to make conclusions.

An example of a rule is 'if a person is male and is married then he receives a married man's tax allowance of 352'. Rules are typically expressed as a list of premises followed by a conclusion, according to the syntax of the system, e.g.:

```
If      (person sex)=male              <condition 1>
and     (person is_married)=true       <condition 2>
Then
        (person tax_code 352)          <conclusion>
```

A premise is merely a test, which requires for its validation an inquiry from the database, e.g. is Fred married? A conclusion is simply a means of adding additional facts to the database, in this case, since Fred is a male and married, then his tax code is 352. In short, rules provide a means of supplying information in specific cases. How does all this relate to the frame system? In the absence of specific facts, rules provide information by concluding results. If I do not know Fred's tax allowance, but I do know that he is married then I can infer it. In the frame system a similar mechanism is available using the if_needed facet: if certain information is not actually present, then it is possible to use a procedure to work it out.

The similarity is the key to the problem of integrating the frame-base and the rules. The answer is quite simply to include in the frame system a statement which says 'if this information is needed, then infer it using a list of relevant rules'. Therefore the inference engine is a procedure which is invoked in exactly the same way as the other procedures used for procedural attachment – via a request for information. Additionally, it takes as an argument a list of rules which are able to conclude a value for the frame's slot, thus:

```
[Fred [tax_code [if_needed [infer [rule1 rule2 .. ruleN ]]]]]
```

But what of the rule-based system's database? In rule-based systems the database is used to store facts about the current state of the problem and the domain, which is exactly what the frame-base does. Thus the rule-based system can use the frame system as a database.

A rule is essentially a list of premises which are tests requiring validation by inquiries from the database, and we have already met the powerful inquiry facilities offered by frames and associative networks. The complex inquiry procedure 'c_inquire' becomes the basis for inquiries of the premises. In summary then, the rules search for evidence in the frame-base, and make conclusions which are also stored in the frame-base. Hopefully the connection of rules and frames should now be clear, and we now move on to the implementation considerations.

At the specification stage of RBFS, the ability to reason with uncertainty was considered essential. This would enable the system to maintain several different hypotheses and at the end conclude each with a given certainty. Another requirement was for explanation both during an investigation and after the conclusions. Since reverse-chaining systems force a directed line of inquiry, they are better than their forward-chaining counterparts for on-line explanation. These requirements suggested a system similar to the famous Mycin system. Indeed, an inference mechanism has been developed which from an external viewpoint is similar, and its features are now described.

8.3.4 Rules, antecedents and predicates

Within RBFS, rules are used to conclude a value for the slot of a frame, and take the general form:

```
IF      antecedent1
AND     antecedent2
AND...
AND     antecedentN
```

```
THEN CONCLUDE
        result with certainty C
```

Premises consist of predicate functions which return a certainty factor as the result of a test on information in the frame databases. The arguments to the predicates are requests for information retrievals which are evaluated using the 'c_inquire' procedure; accordingly the syntax of a retrieval is that of a path (section 8.3.2). There are one and two argument forms for predicates as shown below. Two examples of premises are:

```
Is Fred Married?
True ([[Fred], is_married])

Does the person who Harry loves, love him?
Same ([[[Harry] loves] loves] , [Harry])
```

Largely the same predicate functions exist as in Mycin, but with two additions: Sameclass and Notsameclass. These two very useful predicates exploit the type hierarchies, in order to check for class membership, e.g.:

```
Is any of Fred's pets a dog?
Sameclass ([[Fred] pet],[dog])

Is any of Fred's pets a reptile?
Sameclass ([[Fred] pet],[reptile])
```

8.3.5 Certainty representation and propagation

RBFS uses the same notion of certainty factors as Mycin. There are two types of certainty to consider:

1. Certainty of the evidence, which is a certainty of the data in the information base.
2. Certainty of a rule, which expresses the certainty of the conclusion when the rule has fired.

As with Mycin, certainty is expressed in a range from -1.0 (definitely not true) through 0 (unsure) to 1.0 (definite). The certainty factors are propagated in exactly the same way. The truth of all premises is taken by the minimum of the certainties of the individual premises. If this is greater than 0.2 then the rule fires, and the certainty of the conclusion (as stated in the rule) is modified by the belief in the premises. Multiple conclusions of the same value combine certainty factors using the same combination procedures. Modeling certainty required the frame system to have a means of storing not only the data, but the certainty of the data. This was readily facilitated by storing each value of data as a list, the head being the actual data, the tail containing the certainty.

To model the fact that Fred's hobbies are known to be swimming and more likely jogging, but definitely not squash, then:

```
[Fred [hobbies  [swimming 0.5]
                [jogging 0.8]
                [squash −1.0] ]]
```

To perform predicate tests in Mycin the inquiry mechanism allowed simple access of the form:

```
get (Object, Attribute) → value + certainty
```

In RBFS the 'c_inquire' procedure retrieves values from the leaves of a tree search (as specified by the path). Those values will have an associated certainty, but also the intermediate nodes, on the path to that value, could have an associated certainty. In such cases, the certainties of the values from the search have to be modified to account for the certainty along the path. Since the 'c_inquire' procedure is a successive application of the 'inquire' procedure, essentially it constitutes a series of 'and' clauses; therefore the certainty associated with a value at the end of a path, is the minimum certainty of all the links in the traversal from start node to that value.

8.3.6 Contexts

Mycin had classified the rules which could be applied to certain types of concepts, e.g. rules which applied to organisms. Then for any individual, say organism-1, those rules could be tried. In this case organism-1 was said to be the current context, and Mycin maintained a context tree from which the current context was selected. The notions of segregated rules, general rules and contexts in which they apply were immensely valuable, but with RBFS these ideas fit more naturally in the system, and have been taken a stage further.

Firstly, the segregation of rules is useful because if reduces the amount of work entailed in searching for relevant rules to make a conclusion. Mycin did the segregation for different concepts, but RBFS segregates by concepts and slots. As shown earlier with Fred's tax code, rules are gathered which conclude values for a frame's slot. This means that when a request is made for the values of a frame-slot, if it needs to be inferred then all the rules (and only those rules) which can conclude a value are immediately available in the arguments to the procedure 'infer'. Clearly although the knowledge base may be written in one file it is actually distributed throughout the information model and accessed via:

'if_needed [infer [list of rules]]'.

As for general rules, consider again the example of Fred's tax code. Really the rules apply to all instances of 'adult', therefore in the frame-base they should be attached at the node 'adult', not as shown earlier to Fred. Figure 8.4 shows the syntax of these general rules: the 'CONCLUDE' statement is used to bind the rules automatically to the particular frame-slot, thus:

[adult [tax_code [if_needed [infer [rule1 rule2 rule3]]]]]

An inquiry of Fred's tax code will cause the inquire procedure to look up the 'inst' link to 'adult', and subsequently invoke the procedure 'infer' with an argument of a list of the relevant general rules. But whenever

```
CONCLUDE adult tax_code;

    defrule rule1;
    [  [Same([[^base_frame] gender],[male])]
       [True([[^base_frame]is_married])]
    ⇒ 1.0352]
    endrule;

    defrule rule2;
    [  [Same([[^base_frame] gender],[male])]
       [False([[^base_frame]is_married])]
       ⇒ 1.0222]
    endrule;

    defrule rule3;
    [  [Same([[^base_frame] gender],[female])]
       ⇒ 1.0222]
    endrule;

endCONCLUDE;
```

Figure 8.4 An example knowledge base to conclude a person's tax_code code.

there is a general rule there must also be a means of reconciling it to the current context, otherwise the system will not know to which individual the rules apply.

In the rules given in Figure 8.4, the path inquiries all have start nodes shown as base_frame. This is a variable maintained by the system which contains the name of the frame which originated the request. In fact it is the last name of the specific individual before an inheritance search was undertaken, e.g.:

> c_inquire ([[Fred] tax_code])
> If Fred's tax_code does not exist, set base_frame="Fred",
> and use inheritance to look at 'adult', to see if that can determine the tax_code.

Since the adult tax code slot has an if_needed infer call, the rules, rule1 rule2 rule3, are tried. Suppose rule1 is tried first (actual control will be discussed later), then the premises cause two further inquiries to be issued:

```
c_inquire([[Fred] sex]) →
c_inquire([[Fred] is_married]) →
```

Therefore rule1 will fire concluding that Fred's tax code is 352 (with certainty 1.0).

In Mycin the context tree holds all the concepts of patients, cultures, organisms and drugs, and the current context is given by the position in the tree. In RBFS the context tree is replaced by the much more expressive associative network. The current context is a particular node in the network, and is automatically maintained by the procedure stack of the system. The current value of the local variable 'base_frame' reflects the current context; e.g. suppose that in the context of working out Fred's collateral, we have to work out the wages of Fred's wife (who is Betty), the context moves from Fred to Betty, and after discovering her wages, Fred becomes the context again.

8.3.7 Control and direction

Consider again the example of Fred's tax allowance. The inquiry prompted two further inquiries for his sex and his marital status. If this information is not present in the frame-base, then it needs to be obtained by asking the user. This is achieved with another system procedure 'ask', invoked via procedural attachment:

```
[adult [sex [if_needed [ask[]]]]]
[adult [is_married [if_needed [ask []]]]]
```

But suppose instead that the sex, rather than being asked, could be inferred from some other rules. Then the inquiry would have called infer again, with another list of rules. Thus, the system was trying to infer Fred's tax code using rule1, which inquired Fred's sex, which needs to be inferred from a new list of rules. This is an example of reverse-chaining. Inquiries for information which is not present, stimulates inference (or procedures) to supply the information. These cause further inquiries, and if those data are not present then it causes further inference and so on. The reverse-chaining continues until all the data have been established, usually at the lowest level by asking the user. When the 'infer' procedure gets called with its list of rules to use, there is some scope for deciding which order to try the rules in the list.

In RBFS the rules are sorted into sets for each value that could be concluded. The rules in each set are ordered by the certainty of their conclusions. This is to exploit the 'unity-path' idea of Mycin, once a value is known to be certain (1.0 or −1.0), then further rules concluding the same value can be ignored. When a particular rule is tried it may typically consist of a number of premises, and these could be tested in the order in which they appear in the rule. The first premise may produce an inquiry which itself has to be inferred, or requires that the user answer a question.

However, a later premise may already be known to be false (and therefore enable this rule to be rejected). As with Mycin, the rules are prescanned to check whether they test any facts already known in the framebase, i.e. which can be obtained without having to be asked or inferred. If all the premises which check present information succeed, then the premises requiring further inference are tried to see if these fail the rule. Only if these succeed do the premises requiring questions get executed. This scheme ensures that the user is not plagued with unnecessary questions.

8.3.8 Explanation

The requirement for an IKBS to explain its line of reasoning and justify its conclusions is normally considered essential for the system to earn any credence. Reverse-chaining rule-based systems have demonstrated their competence in this respect: to explain their line of reasoning they simply list the stack of rules which are currently being tried, and to justify conclusions the rules which have fired are recounted. The latter task can be more awkward to implement, since a record must be kept of the rules which fired to produce each conclusion. Whereas for line of reasoning explanation the record of rules being tried is automatically maintained in the stack, in RBFS conclusions are stored in the frame-base as values with certainty, but the third field of the value-list is used to hold the justification of the value:

```
[Fred [should_buy [value [ [house1. 0.7 [rule 19 rule25]] ]]]]
```

The explanation facilities offered allow the user to ask 'why' a question is being asked (explain line of reasoning) and 'how' the system came by a result (justification of conclusions). The explanations may go back to any depth.

8.3.9 Advanced rule concepts

The preceding examples of rules have shown the simple case where rules conclude values (e.g. Fred's tax code), but what if the concluded value is itself a frame' (e.g. that Fred should buy house1 – where house1 is modeled as a frame)? This type of conclusion is very important because it allows relationships between concepts to be established by rules.

There are three cases which arise when concluding frames, dependent on the state of the frame to be concluded:

1. *Connection*: the system can expect the frame to be in existence, so find it and connect to it.
2. *Instantiation*: the system knows that the frame does not already exist, so a new instance of the appropriate type of frame must be made, and connected to.
3. *Conditional instantiation*: the system does not know whether the frame should exist or not. If it does then find it and connect to it; if it does not then make an instance.

These cases of conclusion are handled within RBFS by the inclusion of three special functions; 'a' and 'some' which deal with the instantiation, and 'each' which facilitates the connection. The functions are placed in the 'conclude-value' part of the rule (instead of the simple values), and are executed if the rule fires; they return the names of the appropriate frames.

Connection

The problem here is to determine which frame (or frames) to connect to. Conceptually, to connect a link between two frames, the source and

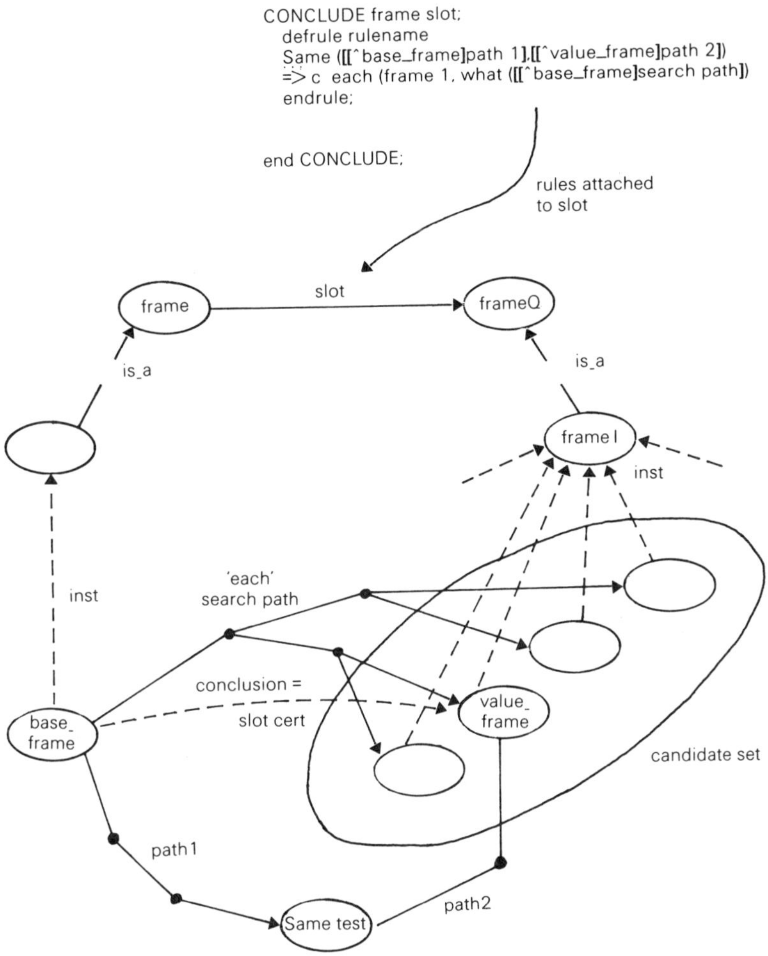

Figure 8.5 The frames involved when concluding frames.

destination frames need to be identified. These are called the base_frame and the value_frame respectively. The base_frame has already been encountered, it is the context of the current inquiry. However, the value_frame needs to be identified. To this end, the 'each' function nominates a set of candidate frames which are considered in turn, by the rule. Figure 8.5 depicts the frames involved in this process.

Consider the house buying example, and a rule which says, 'for each house that the adult is considering to buy, if the adult can afford the house, and he likes the place where the house is located, then he should buy that house'. Which in RBFS is coded as follows:

```
CONCLUDE adult should_buy:
defrule rule 25;
[[Same([[^base_Frame] afford] ,[^value_frame]) ]
[Same([[^base_frame] like_location] ,[[^value_frame] location])
]]
⇒1.0 each ("house", c_inquire([[^base_frame] consider_buy]))
endrule:
end CONCLUDE:
```

Now consider the inquiry, 'what should Fred buy?', which is phrased:

```
c_inquire ([[Fred] should_buy])
```

The system will look at rule25, and initially executes the 'each' function to identify the set of houses under consideration, i.e.

```
c_inquire([[Fred] consider_buy]) → set;
where set= [house2 house3]
```

From that set, it then selects only the items which are of the class 'house' (this would filter out any other objects that the adult might be considering buying e.g. cars, books). Subsequently, for each element in the filtered set, assign its name to the value_frame, and execute the rule. A conclusion will be made for each of the candidate houses, but for houses where no rules fired the certainty of the connection is zero. The scope of the candidate set is defined by the knowledge engineer when he writes the rules. If in the example above, he wanted to consider not only the houses the adult was considering buying, but all houses, then the 'each' part would be written:

```
each ("house",c_inquire([[house] i_inst]))
```

Instantiation

The procedures 'a' and 'some' facilitate the production of new instantiations of frames. However, the nature of the relation represented by the slot may be such that multiple or single values are expected. The 'a' function is used for single values, and 'some' for multiple values. Consider two more rules:

```
CONCLUDE adult spouse:
defrule rule100:
[ [True([[^base_frame] is_married]) ]
```

```
⇒ 1.0 a ("adult")]
endrule:
end CONCLUDE
CONCLUDE adult consider_buy:
defrule rule34:
[
⇒ 1.0 some("house")]
endrule:
end CONCLUDE
```

Rule100 states that if an adult is married then that adult's spouse will be a 'adult', where the 'a' specified a single-valued relation. Rule34 states that an adult could 'consider_buy' some houses. The execution of 'some' will cause the system to ask the user 'how many houses is the adult consider_buying?', to establish the number of frames to create.

The instantiation of a frame is a simple process. The frame is given a unique name (e.g. house9, adult7), and that frame is created with a link stating that it is an 'inst' of the class (which was specified in the arguments to 'a' or 'some'), and finally the conclusion is made linking the base_frame to the value_frame.

Conditional instantiation

When a rule can conclude a frame but it is not known whether that frame has already been instantiated, then the 'a' and 'some' procedures check to see if the frame already exists. In the applications to date this type of conclusion is not common, but absolutely vital when required. It is best illustrated by example. Suppose in the model we want to say the following:

```
An adult lives_in a house
The adult's spouse lives_in a house
```

but with the provision that: The adult lives in a house, and it is the same house that his wife lives in (if it is already known that she lives in a house). When implemented with 'a' and 'some' (without the added conditions), inquiries for the adult and the spouse's house would yield two different houses. Clearly the two need to tie up. In RBFS this problem is overcome with 'instantiation directives', which are lists of information stored in a special facet in the frame, to tell the 'a' and 'some' procedures how to check paths to see if they tie up. If the paths can be tied then select that frame, otherwise make a new instance. There is one other way to tie up frames. The procedures 'a' and 'some' can take a second optional argument which is a flag telling the procedures that rather than inventing a unique name, let the user name the frame. If the user enters a frame which exists, then link to that, otherwise make an instance as before but with the new name. This is termed 'manual' naming and is very useful, e.g. in the house buying example it allows the user to specify that one adult is considering_buying the house which another adult owns.

Figure 8.6 gives a graphical illustration of how information and knowledge interact, using the house buying problem as an example. Figure 8.7 is a schematic of RBFS which illustrates how it can be customized to meet specific areas of expertise. For further details on RBFS the reader should consult Barber *et al.* (1988). For accounts on how RBFS has been applied to a particular test strategy problem in the automatic test equipment industry, refer to Lea (1988) and Garrett (1989). In the next sections we shall be looking at how RBFS has been applied extensively to the area of project management.

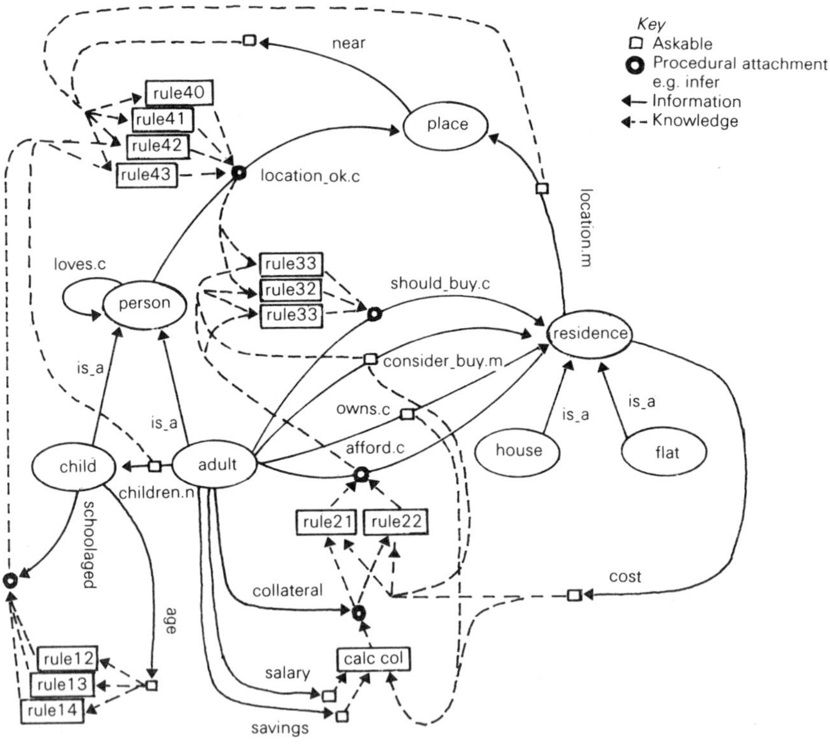

Figure 8.6 The interaction between information and knowledge in the house buying example.

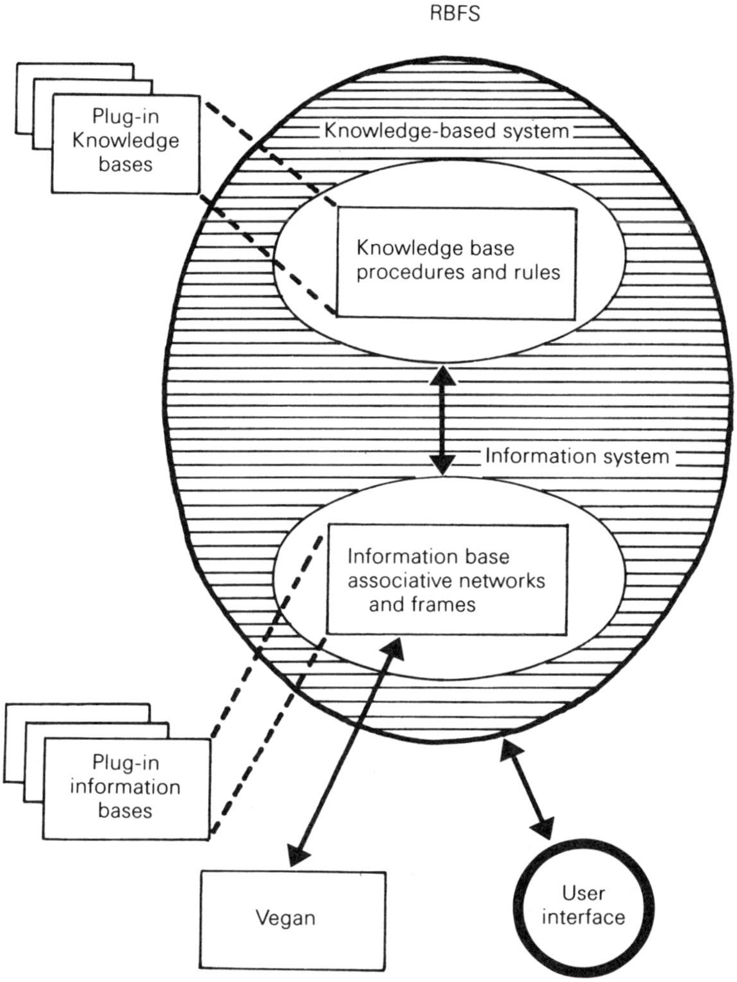

Figure 8.7 The architecture of RBFS.

8.4 SYSTEMS ENGINEERING AND PLANNING

The reader will recall from Chapter 1 that the definition for systems engineering included the term 'planning'. We turn now to a more detailed consideration of this aspect of engineering for two chief reasons. First, planning is an essential element of the systems engineer's responsibilities. Secondly, the author's personal research with RBFS has included the development of expert systems for planning tasks, particularly autonomous plan generation and project control (or expedition). In section 8.6, case studies for the application of these expert systems are reported at some length. In this section we consider the following:

1. One particular interpretation that engineers have put on to planning, namely project management.
2. The failure of project networking techniques as a methodology for effective project management.
3. A re-examination of the intellectual enterprise which planning clearly represents, as a prelude to producing an improved methodology for project management that takes advantage of the expert system paradigms described above.

8.4.1 Planning and project management

A perfunctory reflection of the reader's own personal experiences with planning may result in thoughts of taking a holiday, redecorating the house, making an investment on the stock market, or even just simply getting up in a morning. The engineer's need to plan, whilst being set on a more professional basis than any of these domestic examples suggest, is no different conceptually. The major difference is that so far as the engineer is concerned, it is an *imperative* exercise.

Businessmen and industrialists the world over when planning a new facility are faced with the task of selecting the most economical and effective method of executing the *project*. In many instances, the work involved in translating the facility from a concept into an operable and profitable reality is so complex as to require specialist expertise in the planning function. Where this is lacking within the organization requiring the new facility, it can be found by employing professionals whose business is primarily concerned with project management.

A project is generally regarded as consisting of discrete phases. The first of these is project *definition*. Effort spent in this phase gives major paybacks in the subsequent phase, project *execution*, in terms of the smoothness of performance and economical factors. Project definition is in essence a detailed requirements analysis, which establishes the key technical and economic objectives for the project execution phase. It usually results in a project execution plan covering the following:

1. The procedures for the project.
2. Design objectives, criteria and standards.
3. Definitive project cost parameters.

4. The project control schedule.
5. The project procurement and subcontract plan.
6. All local authority and utility requirements.

The motivation in the project execution phase is always towards achieving the necessary quality at the minimum cost and in strict accordance with stated deadlines. Execution necessarily proceeds on a day-to-day basis whilst having to provide reports on progress at regular intervals. Such reports will cover the following:

1. Provide an executive overview.
2. Refer to the milestones achieved.
3. Highlight impending problems and propose solutions.
4. State the key objectives next in line.
5. Give a financial status.
6. Review progress on engineering design, procurement and construction.

Progress reviews on engineering design, procurement and construction are worthy of special mention. The construction of many new facilities such as a coke-manufacturing plant is a complex operation calling for the detailed design of vessels, sophisticated piping and plant layout. Often the plant will require the incorporation of state-of-the-art electronics for control, shutdown, security and monitoring systems. All this engineering design is better done in an integrated fashion since anomalies occurring in the overall configuration can seriously disrupt the smooth running of a large-scale project. Procurement involves purchasing and expediting equipment to ensure delivery on-time of the appropriate assemblies and components. It further requires quality control surveillance of suppliers to ensure a reliable source of supply, and that the shipped equipment arrives in timely fashion in 'ex-works' condition.

Effective project execution will usually mean concurrent design and construction, thereby enabling early completion of the project. Such concurrency is usually only possible by means of sophisticated computing tools for both design and scheduling, and even then there must be clear policies for the subdivision of construction work into discrete and coherent packages. Site activity is an example of concurrency, where the design of advanced electronics equipment needed for plant commissioning may proceed in parallel with the erection of structural steelwork for the installation.

It therefore requires little imagination to envisage the complexities of project management, the opportunities for catastrophic failure and the need for resourceful inventiveness to keep the whole under control.

8.4.2 Computer-based project management tools

It is a remarkable tribute to the engineer's genius that he has in any sense been able to develop techniques which in a formal way help to prevent a complicated interaction of elements from exploding without warning into complete anarchy of cataclysmic proportions. To this end project networking techniques (PNTs) have been developed alongside a number of their derivatives for describing the status of the project underway. Section 7.6 provides an outline description of one such PNT. The representation used there was activity-on-link (AOL). An alternative representation is activity-on-node (AON), which is used to report on the results of some expert systems research given in Section 8.6. Figure 8.8 gives an AON representation for a typical project network.

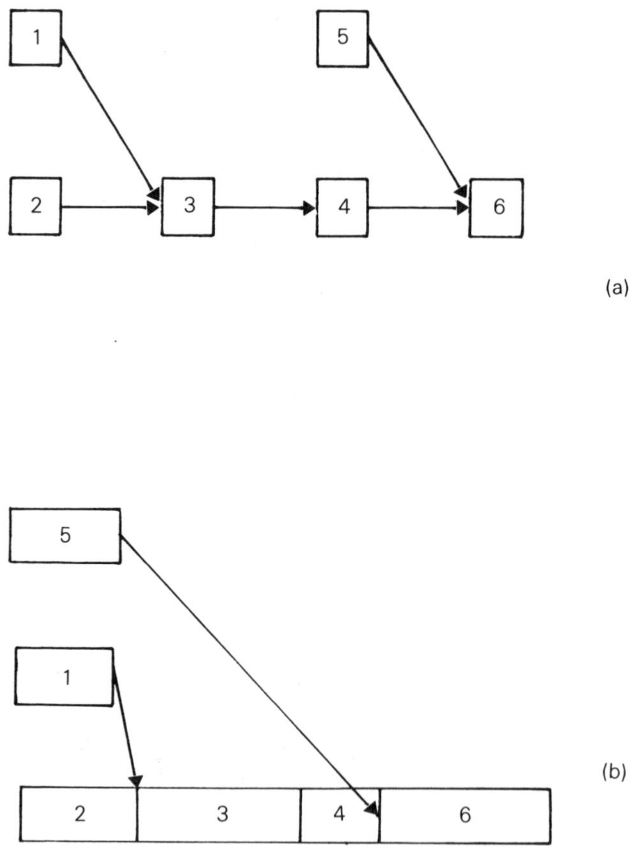

Figure 8.8 (a) Activity-on-node (AON) format. (b) Time-scaled AON format.

PNTs began to emerge in the 1950s much to the fascination and pleasure of the planning fraternity. Its members had been struggling with the problem of developing a basic logic to synthesize and analyze 'the processes by which stated aims and objectives could be realized by practicable means'. PNTs did the following:

1. Pictorialized a logic to enable planners to see what they were up against.
2. Provided a disciplined framework for activity sequencing, statusing, resource allocation and scheduling.
3. Offered a useful means of instruction to would-be planners.
4. Showed themselves amenable to computer processing.

Having arrived on the scene they were considered to be 'indispensable', even though they were 'not without their faults'. These 'drawbacks' have not diminished in the intervening period, rather they have worsened as planners have been given more ambitious schemes to contemplate and even tighter constraints in which to fulfil them. It is not uncommon today in large-scale projects for networks to be developed comprising tens of thousands of activities. Not surprisingly it has been observed that the management problem has been transferred from the engineering of the plant to the administration of the network. Moreover, advances in computing have failed to stop the salient and inherent difficulties that the PNTs possess. Stated briefly, these appear to center on the following issues:

1. Maintenance of the large-scale network.
2. Abstraction of the network for different levels of management.
3. Isolation of the network from essential detail.
4. Impoverishment of knowledge contained by the network.

There is undoubtedly something essentially wrong with the PNTs, and there is little cause for optimism that the next generation of computers will 'crack' the basic difficulties. Before any progress can be made in getting to a solution, a complete reappraisal must first be carried out of what kinds of planning logic and visualizations are needed.

Not for the first time is the engineer confronted with a *model* whose regulation poses an equal degree of perplexity as the *reality* which the model seeks to describe. In the author's view, the systems engineer is the one who can design *systems* in which the 'reality component' and the 'model component' are not in competition, each apparently working for the downfall of the other, but are rather mutually supportive. How this might be achieved, taking as our testbed the problem of project management, will require us first to examine the essential intellectual enterprise which constitutes the planning process.

8.4.3 An intellectual enterprise

Planning means different things to different people and yet something to everybody. This points to it being a generic intellectual exercise. People make plans in a multitude of ways and they execute them with equal variability. This would indicate that a number of levels of reasoning strategies are brought to bear in the intellectual effort that planning consumes. Finally, people learn to plan through failure. It is important to recognize the influence that this learning has, for this ought to be not on the next plan, but rather on the planning mechanism for its generation and control. Getting the plan right should be seen more in terms of the reward for an effective planning mechanism, rather than as a prerequisite to the success of the plan's objective. In arriving at a satisfactory definition of the 'planning problem', it is vital to separate the *message* of planning (contained in its generic structure, its levels of reasoning strategies, and the learning component) from the *medium* of planning (messed up by the plethora of planning tools).

The task confronting the project manager is essentially that of maintaining a delicate balance of a multitude of interrelated variables to ensure expeditious project execution. A project which takes no account of feedback responds to every undesirable influence, with a resultant increase in either project time-scale or cost. The project manager is constantly seeking to exert his control in order to 'dampen' these influences. The nature of the profession, therefore, is one of 'seat-of-the-pants' management, based on years of experience accumulated in the project manager's highly specialized field.

The questions we must ask are: 'what are the essential components of the engineer's experience?' and 'what are the fundamental intellectual constructs involved in arriving at an effective control scheme for the administration of a project?' If we find satisfactory answers to these then we stand a chance of developing systems, possibly ones which a computer can be made to understand, which will ease the administrative burden, thereby enabling projects to be completed more satisfactorily and ensuring that the engineer gets his job done properly.

Two elements at least are of striking importance. First, in a project, many different perspectives are held by those occupying various positions within a naturally occurring hierarchy. Classically there is the problem-owner – the party with a need, e.g. to have a new plant built. Then there is the contractor – the party who offers, for a fee, to solve the problem-owner's problem for him. There usually appears on the scene a consultant, retained by the problem-owner to ensure that the problem-solver keeps to his offer. Endless disagreement between contractor and consultant, started off to be primarily in the best interests of the problem-owner, can be counter-productive. Each of these primary parties will have their own functional subordinates who will become involved at appropriate times as their expertise best matches the prevailing difficulties which obtain throughout the various stages of a project's life cycle. The inevitable conflict of differing *viewpoints* ensues.

A second key element necessary for the understanding of the 'anatomy' of a project is that of judgment. Projects are all about value-judgments and mechanisms which resolve such issues as what to do first, who should be assigned to which task and where the greatest need is for these resources. Decisions of this generic kind require reasoning with such concepts as concurrency, interrupt and priority. These decisions must be made with incomplete information, perhaps based on misleading information, with uncertain outcomes, under pressure and with little time available before it becomes meaningless to make them. Who would be a project manager? As it has been so crisply stated:

> project management is no easy task. The unforeseen happens and things go wrong. People fall ill, machines break down, and the money runs out. Goals are glossed over, and milestones are missed. People get frustrated, angry, even irrational. Heads roll while the programme doesn't.

In the following section a system is described that has been designed to deal with this problematique, using some of the expert system paradigms described in the previous sections. Maybe artificial intelligence can achieve what the real variety is not up to!

8.5 AN AI METHODOLOGY FOR PROJECT PLANNING

The reader will recall the remarks made in section 2.5 regarding methodology. There it was said that methodology is not method. The word should convey a higher level of intellectual enterprise dealing more with a set of principles of method. It should however provide a framework in which precise models, techniques and concepts can be organized so as to enable its prospective users to achieve a greater degree of control over problematical situations than a model or technique itself actually allows.

An AI approach to project management is presented in this section that uses expert systems as a means of formalizing and embodying the informal, empirical techniques of management. This has required a scheme to be developed for representing information about a project; information with which intelligent reasoning can be conducted. The methodology stems from a recognition that current project modeling techniques, such as Pert-type networks or Gantt bar charts are grossly inadequate for portraying detailed information suitable for these reasoning processes. The elements of the methodology are threefold:

1. An original and powerful project modeling schema which will allow detailed project definition.
2. Conceptual sectionalization of an information base which facilitates efficient and intelligent project description.
3. A distinction between knowledge and information.

Attempts to model a project necessitate the initial development of a model for an activity within the project. A simple model is to consider an

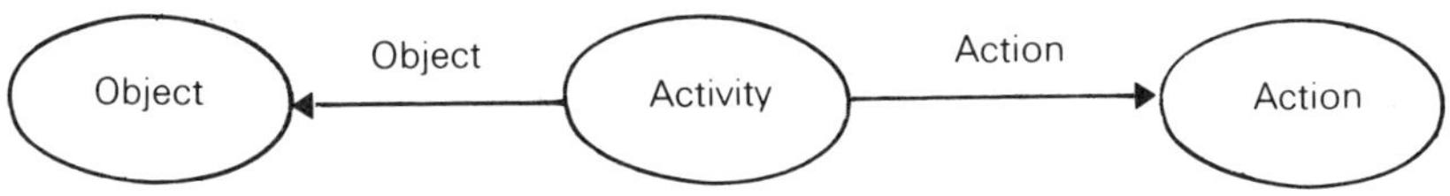

Figure 8.9 Activity defined as action on object.

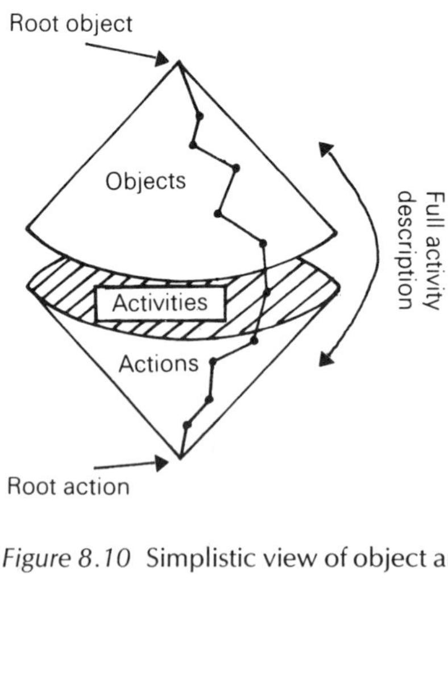

Figure 8.10 Simplistic view of object and action hierarchies.

activity to be the application of an action to an object; Figure 8.9. This structure is analogous to that of an English sentence where the action is the verb, and the object is the noun. This representation may easily be expanded by recognizing that each object may be a component of a larger object, and each action merely a composite part of a grander action. This structural form is vividly portrayed as two cones attached by their bases, each cone representing the object- and action-hierarchies respectively; Figure 8.10. In this case the attachment plane would represent the activities themselves described in their most recognizable form, e.g. 'perform electrical test no. 23 on printed circuit board no. 4'.

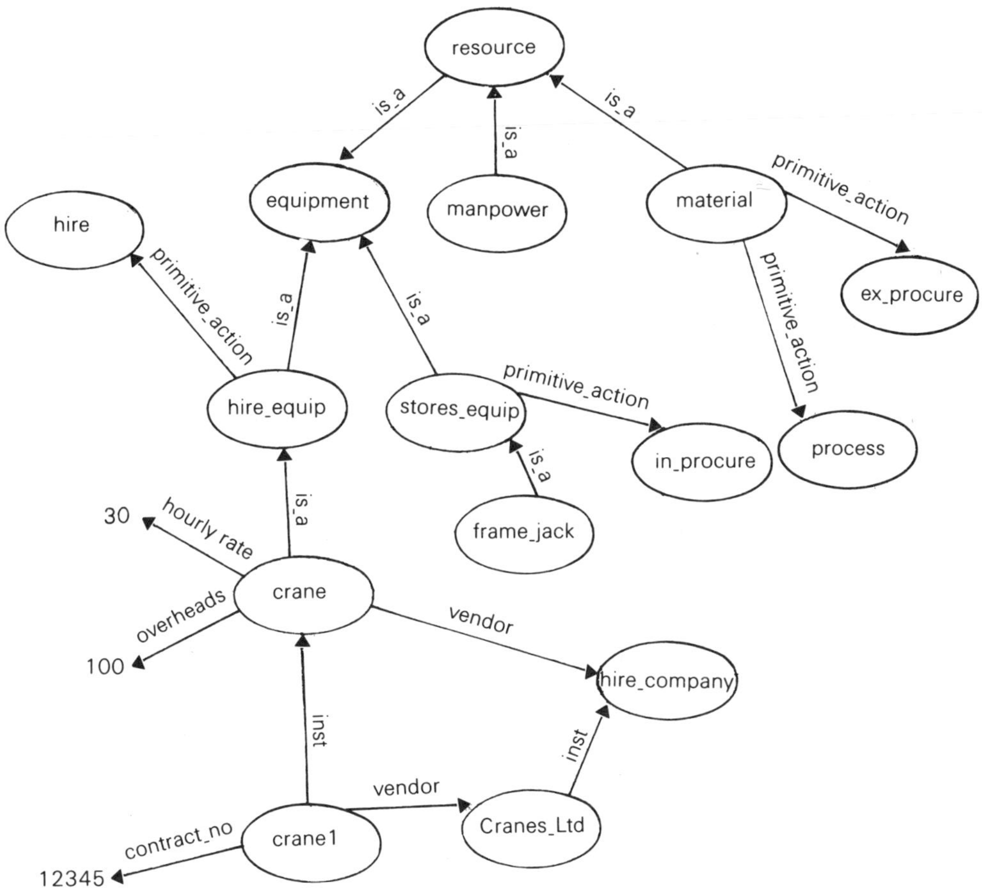

Figure 8.11 Resource class hierarchy.

```
(resource (i_is_a (value (equipment)
   (manpower)
   (materials)))
(cost (if_needed ((calc_resource_cost ()))))))

(equipment (is_a (value (resource)))
   (i_is_a (value (hire_equipment)
      (stores_equipment))))

(hire_equipment (is_a (value (equipment)))
   (i_is_a (value (crane)))
   (primitive_action (value (hire)))
   (contract_number (if_needed ((ask ())))
      (range (number))))

(crane (is_a (value (hire_equipment)))
   (i_inst (value (crane_1)))
   (overheads (value (100)))
   (hourly_rate (value (30)))
   (vendor (value (hire-company))))

(crane_1 (inst (value (crane)))
   (contract_number (value (12345)))
   (vendor (value (hire-all-cranes-ltd))))
```

Figure 8.12 Resource class hierarchy frames.

Three simple rules to define the construction of project models have evolved:

1. An activity is considered to be an action on an object, the particular action and object have to be identified for each activity, and drawn using the notation of activity node with 'object' and 'action' links.
2. The objects form a hierarchy. The 'highest' object is shown connected to its subcomponents via 'sub' links. An object in the hierarchy can be either physical or conceptual.
3. The actions form a hierarchy in similar fashion to the objects.

The project model described thus can be represented visually as an associative network, whilst it can be stored internally as a frame-slot schema, both 'devices' being familiar in AI research; Figures 8.11 and 8.12.

This hierarchical definition of the project, facilitated by the simple

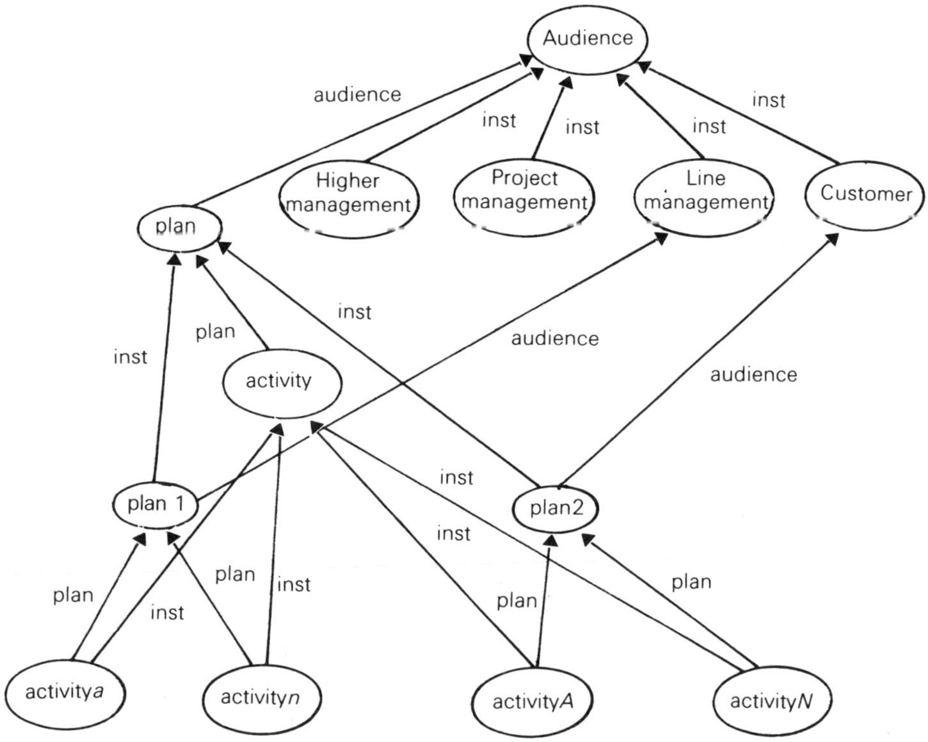

Figure 8.13 Plans of various abstractions may exist in a single model, the activities of each associated with different 'plan' nodes having alternative audiences.

acceptance of an activity being constituted as 'action' upon 'object', provides a basis for a richer and more powerful representation of a project which is truly wealthy in information, dwelling in what AI researchers can only describe as a 'rich environment'.

The extension of this elementary model is afforded by four further features:

1. Activity abstraction.
2. Attribute attachment.
3. Class hierarchies.
4. Precedence linkage.

Activity abstraction is modeled easily by recognizing that certain portions of the 'cone model' form closed networks. Thus for a given

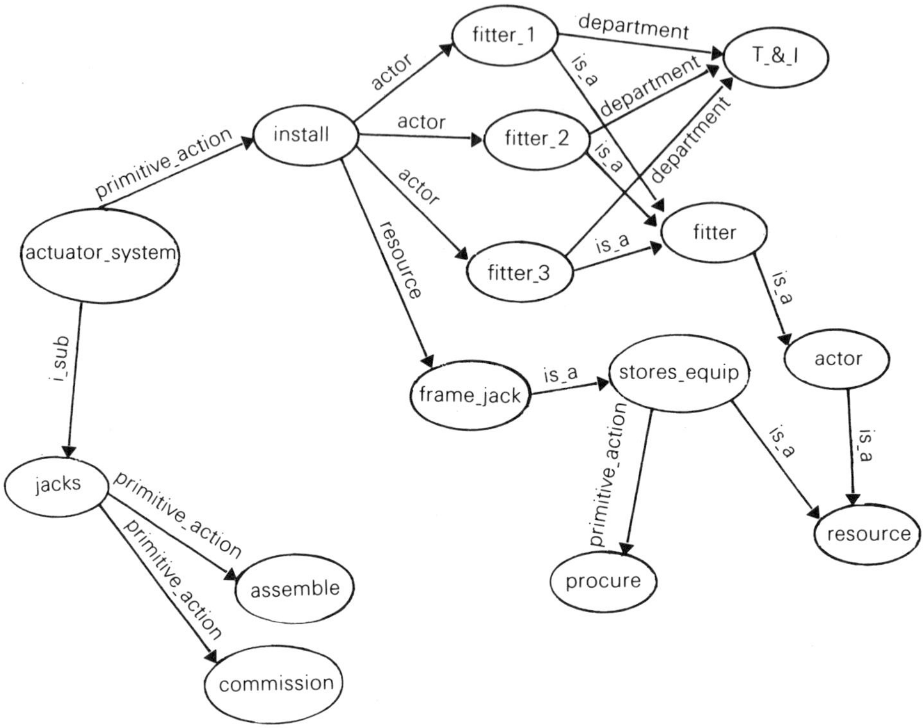

Figure 8.14 Actions require resources, resources may have actions themselves.

object–action pairing it is possible to 'collapse' this sub-network into an abstracted activity thereby providing a means of delivering information of an appropriate kind to a given audience, Figure 8.13. The reader need not be concerned that the detail in Figure 8.11 is obscure in places. First, Figure 8.12 gives this detail in a different representational form. Second, it is of greater importance that the reader appreciate the overall visualisation of project models represented as associative networks. The ability to spot emergent patterns is a skill which the systems engineer should develop. Facilities exist for the user of the expert system to pan and zoom around the entire associative network in order to inspect the details at any desired level. Figure 8.11 and others like it in this chapter, have been reproduced from actual computer generated images that are naturally supplied by the expert system in the course of its use.

Attribute attachment is an obvious extension upon inspection of the associative network form of the project model. Thus, to any of the actions may be attached the attributes of say 'actor', 'resource', 'location' or some other such piece of refining detail. Indeed, to any of these attributes, which are entities in their own right, e.g. 'actor', further attributes may be attached, such as 'rate_of_pay'. Thus the fine detail of the associative network unfolds as more of the 'environmental' information affecting project behavior becomes part of the project model; Figure 8.14.

Class hierarchies are a means of building efficiency into the informational representation. Using generic concepts, instances of which figure in detail both in the action- and object-hierarchies, information

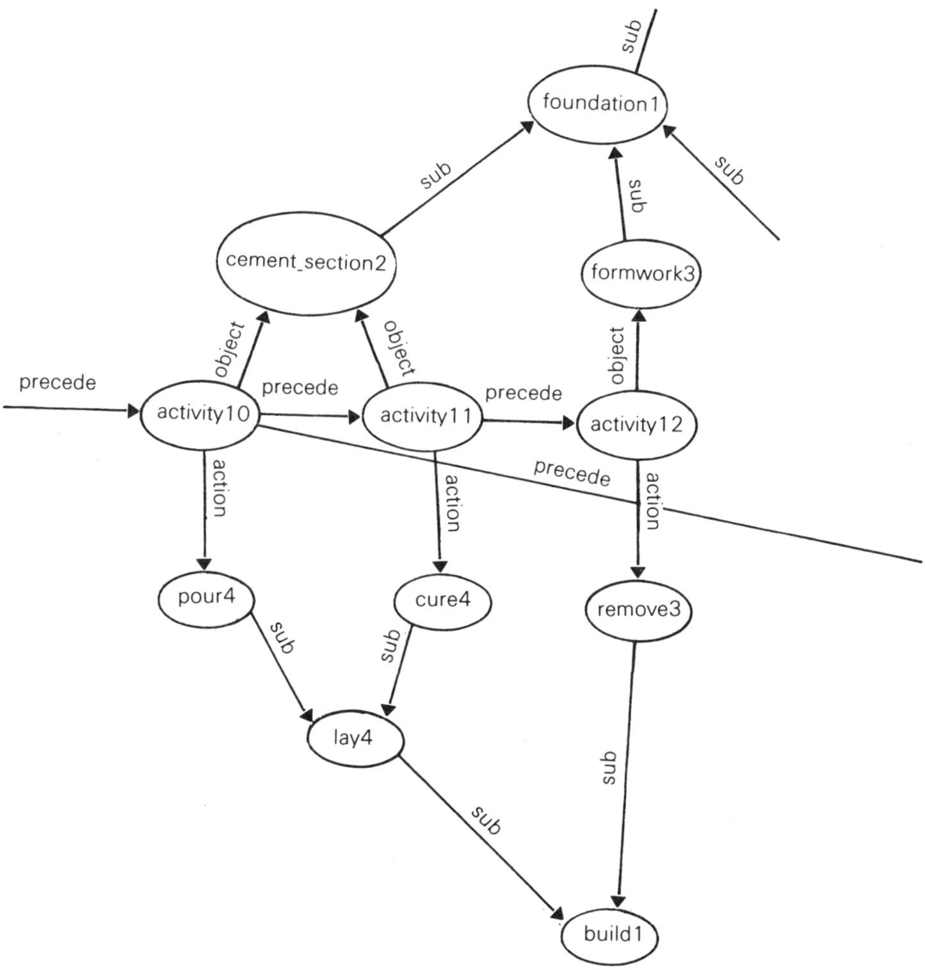

Figure 8.15 The precede link to specify logical dependencies between activities.

common to these instances may be attached to the generic concepts and thereafter retrieved by referral to the generic concepts from the instances via 'emotive' links such as 'inst' and 'is_a'; Figure 8.11.

Precedence linkage, the final extension to the project model, shows how the model relates to the original project network in which there must be found a sense of sequence or precedence. A 'precede' link, attached to the activity nodes is a convenient means of defining the sequence in which activities, at their 'base' level are to be executed. With this link the model demonstrates the complete project network in its AON format. In

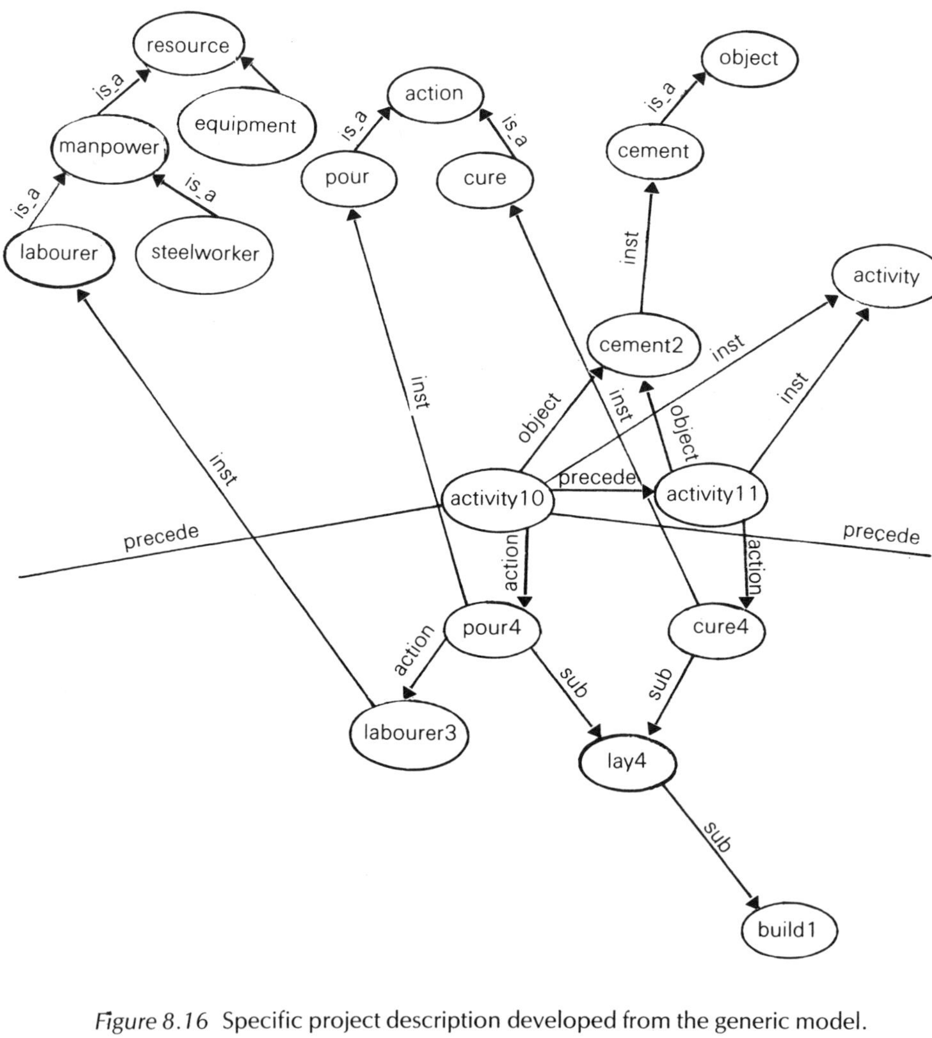

Figure 8.16 Specific project description developed from the generic model.

effect the project network is now revealed as simply a single dimension of what in truth has always been a multi-dimensional scenario, a fact dramatically underscored by the associative network model, Figure 8.15.

Thus far, the information about a particular object has been described in terms of the objects, actions, actors and other relevant entities which can be described as project-specific. However no project is an island, and this is vividly demonstrated by the wealth of information which a project manager draws upon to formulate and execute his plans. The project environment must also be represented in the information base and the methodology provides for this by making a conceptual sectionalization into general and specific planes. The latter has been adequately described already, while the former is worthy of further remarks.

The general plane will include all possible modules that can be included within the overall product that the project seeks to engineer, and knowledge about how to decide whether anything specifically should be included in the product with which the contemporary consultation with the expert system is concerned. Additionally the general plane furnishes details of how to schedule activities that are found to be part of the project. It is a statement from the project managers of all that could possibly be entailed in product definition. It will comprise, among others, information on and knowledge concerning contractors, authorities, clients and the host company (including such data as personnel types, departmental organization and product history); Figure 8.16. For further details on this methodology refer to Marshall *et al.* (1987a). For accounts on the extensive experience gained with this methodology consult Boardman 1988, 1989a, 1989b, Boardman *et al.* 1988a, 1988b.

8.6 EXPERT PLANNING SYSTEMS

An expert planning system can be defined as an automated reasoning tool which implements expert knowledge of constraints, conflicts, global goals, time-lining and heuristics of a planning situation. The system must assist its users to solve planning problems by reasoning with the possibilities of the implemented expert knowledge.

There is an obvious analogy between expert systems and expert planning systems: both implement expert knowledge of a limited domain, and perform some automated reasoning to assist the user. The power and usefulness of both types of system also depend on the quality and quantity of the expert knowledge implemented, and development therefore requires close collaboration with the domain experts. Expert systems use rules to derive new facts from known facts, whilst expert planning systems use plans to derive new goals from existing goals.

The analogy however extends no further than this. Expert planning systems require extensive representation of planning constraints such as resources and time. Further, the reasoning within an expert planning system requires complex side-effects to be taken into account during the reasoning process. It is generally desired that plans should be optimal in some sense, e.g. minimal resources, minimum total time, plan flexibility. Since most real planning problems are known to be NP-hard, owing to the large solution search space, expert planning systems generally also need to implement some heuristics in order to search efficiently for adequate, sub-optimal solutions of the planning problem. Very few expert planning systems have the capability for multiple lines of reasoning or an adequate concept of time.

Reasoning and knowledge representation schemes that are much more sophisticated than those currently available are required to implement an expert planning system for expert system development. The author has been involved in a program of research in association with two industrial collaborators, in which particular project management problems have been identified: those of bid proposal and project expedition. The research program attempts to satisfy real-world problematic issues by producing expert systems that tackle these situations in a pragmatic way. These systems incorporate the results of considerable (and ongoing) knowledge engineering, conducted in collaboration with Rediffusion Simulation and Babcock Contractors respectively.

8.6.1 PIPPA

The first system, called the Professional Intelligent Project Planning Assistant or PIPPA (Marshall *et al.* 1987b, Kellett and Boardman 1989), addresses the problem of bid proposal in a large engineering company – Rediffusion Simulation Ltd (RSL). The company is a manufacturer of flight simulators for commercial and military aviation clients. The bid proposal problem is one of speedily assessing a client's requirements, and, subsequently, identifying time, cost, effort and risk data which can be used to bid for the contract to produce the simulator. The salient aspects of this process are as follows:

1. Identifying the features of the simulator that satisfies the client's needs.

2. Producing project plans (in the form of project networks and bar charts) to convey the time-scale of the project and to identify activities which constitute the project.

With plans produced, reasonable time and cost estimates for the project can be given to the client for assessment. To produce a reasonable bid requires a full assessment of the complexity of the flight

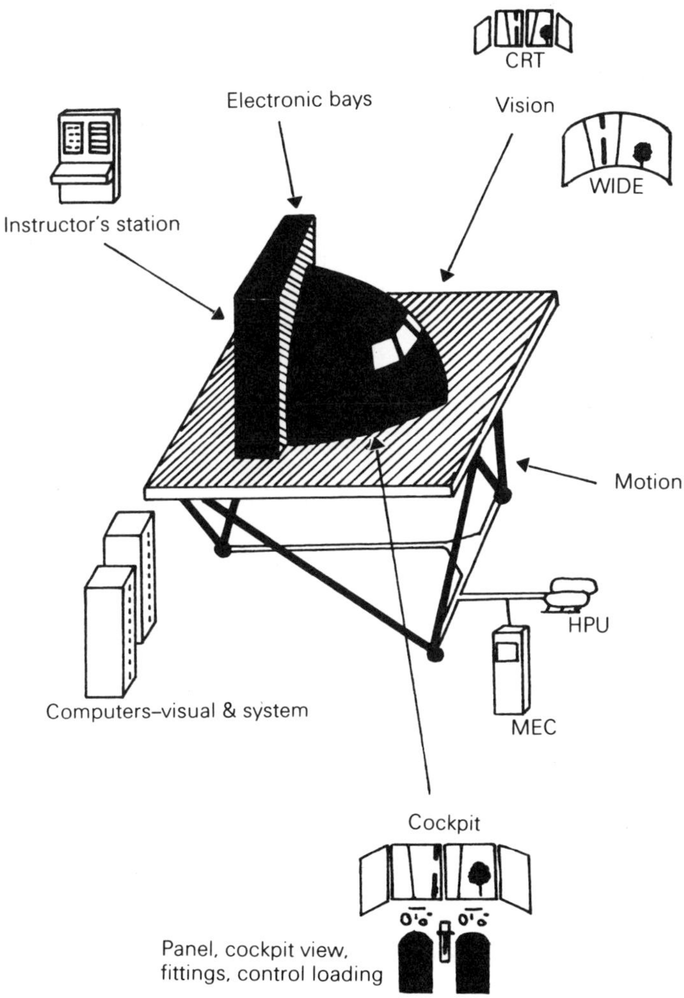

Figure 8.17 Schematic of a flight simulator.

Airport terrain models:	no. of airports specific airport replication variable terrain detail generic airport model
Vision projection	CRT: no. of displays no. of channels WIDE: angle of display no. of channels
Vision software model:	night flight dusk/night flight day/dusk/night flight variable textural detail
Computers:	type configuration latency spare capacity
Instructor's station:	no. of stations panel layout malfunctions lessons record/replay
Special effects:	fog/rain/snow turbulence – in-flight/ground fire/smoke
Motion:	jack length HPU cooling plumbing
Cockpit equipment:	control loading panels instrumentation fittings
Software:	flight characteristics

Figure 8.18 Simulator configuration possibilities.

Pilot category
FAA-I: Fully qualified air carrier pilots.
FAA-II: Highly experienced pilots (at least second in command of an airplane of the same group).
FAA-III: Possess an airline transport pilot certificate + 1500 hours flight time.

Training objectives
FAA-I: current landing maneuver standard.
FAA-II: Transition (between aircraft of the same group) and upgrade training.
FAA-III: Total initial, transition, upgrade or recurrent training.

Figure 8.19 Federal Aviation Authority phase approvals.

simulator, and thereafter to derive an accurate match of product to requirement specification. The pressures on the company (RSL) to achieve this on time and with minimal expense are all too obvious. A flight simulator is a sophisticated piece of equipment which simulates a particular aircraft. Through the replication of the flight deck and the aircraft's responses (motion), and the simulation of characteristic cockpit views (vision), a range of taxing flight lessons may be undertaken (instruction). Aircraft take-offs and landings at specific airports, in-flights fires and turbulence may all be simulated. Figures 8.17 and 8.18 present the typical generic configuration of a simulator, and a partial inventory of components and their use. Figure 8.19 details the potential scope for sophistication.

Each simulator manufactured is unique in that it is an exact replica of a specific aircraft owned by the customer (i.e. the simulator is made of a certain tail-numbered aircraft). Additionally, every simulator is slightly different in its level of sophistication.

In order to win a contract with a customer, RSL will have to compile an estimate of product structure, cost and project time-scale to be submitted, along with the competition, to the customer. Unfortunately bid compilation is a time-consuming task that may take between 4 and 16 weeks to complete. In the assessment of such a bid, the company must consider the client involved, the complexity level required of the simulator, the novelty of any simulator components, the detail of the client's specification, along with countless other factors. Figure 8.20 presents a resumé of some of these. Obviously, the bid must be assessed fully enough so that any peculiarities that exist are appreciated for their project impact and cost. Conversely one does not wish to waste too much time or money on compilation of a bid where success is uncertain. A trade-off is inevitable.

One, rather exhaustive, evaluation strategy is to employ the talents of a project manager for every bid compilation to be made. In conjunction with the line-managers of the company, the prospective simulator's specification may be investigated thoroughly for its product and project import. Alternatively, as is currently the case, a sales engineer may be called upon to compare the target specification with the specification of simulators for which past bids have been compiled. By observing the commonality between the currently proposed simulator and a range of previous ones, the sales engineer hopes to draw together the list of simulator components and project activities implicated. In this way, the sales engineer hopes to speed up the compilation process by drawing from previous bid data packages the common data items – making necessary cost adjustments in lieu of inflation and the like. If a bid is to involve many new features, then a project manager may be assigned to the proposal to investigate further.

Authority involvement: Implicit requirements are associated with various authorities. This may require that particular test equipment be incorporated, certain components be included, and minimum levels of sophistication and complexity (capability) be utilized.

Third party involvement: Certain major components may be manufactured by third parties – competitors. The inclusion of such equipment onto the simulator might require modifications to be made.

Customer specification: The customer may explicitly ask for particular levels of equipment; project restrictions may be imposed on time, cost or control mechanisms; the delivery environment might impose restrictions on which components can be included; the country of destination may require special investigation.

Aircraft type: The specification of the aircraft to be simulated will obviously impact upon the cockpit fittings required, but will also affect the software required to effect simulation.

Figure 8.20 Typical information to assess simulator configuration.

So, for example, if the current simulator is to have a particular visual projection mechanism, one must correctly identify the explicit components (projectors, drivers) as well as the implicit components such as mounting brackets and containers. This example shows the need for a formalism to be employed in the description of component necessity and justification. Simply looking at previous configurations will not be adequate unless sufficient documentation has been compiled so as to describe both implicit and explicit necessities.

In the expert system proposed in solution to this problem, the configuration information and knowledge implicit in this historical data is distilled into discrete justifications for individual components or sets. By describing the library of components from which a flight simulator may be compiled, and the manner in which these components may interact, the information and knowledge may be modularly expanded and efficiently maintained. This solution, based upon the assimilation of historical data, works because, although each simulator is unique, it can be simplistically viewed as conforming to the same underlying structure; its components largely pulled from a standard range. In other words, a simulator almost always has a vision system, a motion system, a flight deck, an instructor's station, a computer installation ... and most of these components are drawn from the standard range offered by RSL (there is a range of three visual display systems, six image generation systems, etc.). Each of these simulator subcomponents are themselves compiled from a fairly standard range of components.

The aim of the expert system proposed is to provide a formalized approach to the definition of a simulator during this bid phase. Calling upon the project manager's and sales engineer's knowledge and information, the proposed system will decide upon the configuration that the simulator should take, and the manner in which the resulting configuration should be constructed. In order to describe how the simulator might be constructed, one must assess the likely activities that will be performed. Since a configuration will have been identified, many of the actions might be obvious: particular actions required to obtain or manufacture the constituent parts. But others might be slightly more obscure. For example, certain integration tests may only be required for some configurations. Customer liaison activities may be incorporated, particularly if the military is involved. In terms of results, the system should provide detailed data which describe the configuration, cost and project networks, and bar charts of the project which could be used for presentation to line managers, project managers, higher management and the customer.

Figure 8.21 shows a generic model of project planning which PIPPA utilizes to instantiate specific plans, and Figure 8.22 shows the generic rules by which this instantiation process is governed.

8.6.2 XPERT

The second system, eXpert Project Expedition Reasoning Toolkit or XPERT (Barber and Boardman 1988a, 1988b), addresses the problem of project expedition, and this work benefited from extensive collaboration with Babcock Contractors (formerly Babcock Woodall-Duckham), a large company with an international reputation in project management.

Expedition is the process of reducing the total duration of a project towards a particular target date. The necessity for expedition arises when a project or individual activity is in danger of running grossly over time, and this has to be curbed in order to avoid project penalties or company credibility. To expedite a complete project it is necessary to identify a number of activities within a project which would be most suitable for expedition and which will have a considerable ameliorative effect on

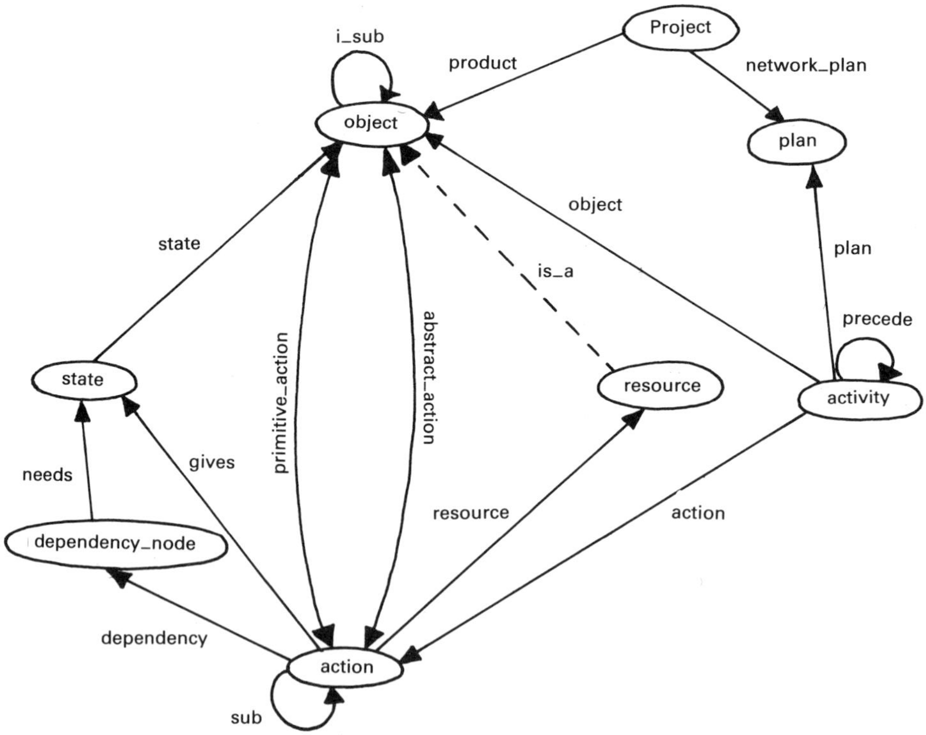

Figure 8.21 Generic project model.

Rule 1: A project is defined if its product (an object) has been defined and planned.

Rule 2: An object is defined if its subcomponents have been defined and its primitive actions and abstract actions have been defined.

Rule 3: An action is defined if its component relationship has been evaluated (super-actions) and its resources are defined.

Rule 4: An object is planned if its action dependencies are assessed and its activities have been identified and sequenced.

Figure 8.22 Generic planning rules.

meeting the desired project time-scale. Expedition invariably incurs some increase in direct expenditure for an activity, as money is spent to save time, for example, by the deployment of additional resources. The aim of the expedition process is to meet the target date, but with a minimal increase in total project cost. The expedition process, therefore, is one of optimization of time against cost.

The systems approach to the expedition problem adopts a holistic viewpoint which expands the problem from just an algorithmic solution, to encompass the problems associated with gathering the data, the latter problem being at least as difficult (in a different way) as the traditional optimization problem. The starting point for the algorithm is the data that it requires; namely the CPA network and the time–cost trade-off function for each activity. The former is readily available, the latter difficult and time-consuming to postulate. Working backwards, the only way to determine the time–cost trade-off function is to know the particular strategy of expedition which is to be applied. Hence, the first step towards automating the data collection process is to be able to postulate an expedition strategy for an activity, and to estimate the associated time–cost trade-off function. These two processes are now discussed and the model on which the analysis is based is presented.

A perfunctory analysis of the potential methods to expedite activities reveals a hierarchy of abstract expedition strategies; Figure 8.23. A brief explanation of each strategy now follows. Some of the factors for the suitability of different strategies, in different situations, are presented later.

The first area, *more control*, is about improving the performance of the workforce: incentive schemes in this context can provide a climate for increased productivity in labor-intensive environments. Improved organization is an aspect of good management and leads to improved efficiency.

Finally, closer monitoring ensures that workers are actually attaining a reasonable level of performance; in particular that they are not under-achieving.

More time is a strategy for obtaining more time to work on an activity in the same amount of project elapsed time. The two methods are overtime and shift work, both of which pay premium rates. There is however a limit on the number of hours that can be worked on overtime both legally and physically, and consequently, there is a limit to the amount of time saved on an activity. The same is true of shift work, although more time can be saved because it is possible to work 24 hours a day. An added complication is that shift work requires extra manpower to populate each shift, and in some cases there can be problems handing over work between shifts.

The strategy of *resources* suggests that extra resources can be deployed to assist and thereby expedite the task. There are three subtypes, 'manpower', 'equipment' and 'materials', but there are practical limitations to the amount which can be added. For example, only so many laborers can excavate a hole, depending upon its size.

Change contract applies in situations where a work chunk has already been contractually agreed with a third party. Examples are of subcontract work and certain purchase orders. In some cases it is possible to renegotiate to achieve tighter time-scales, but this will inevitably cost more. One crucial factor is how far the project is into the contract. If the contract work has not begun, then there is a greater chance of successful renegotiation.

Changing the specification of an object might enable an activity to be more easily and quickly achieved. An example is where there is difficulty obtaining the specified size steel bar to go into reinforced concrete;

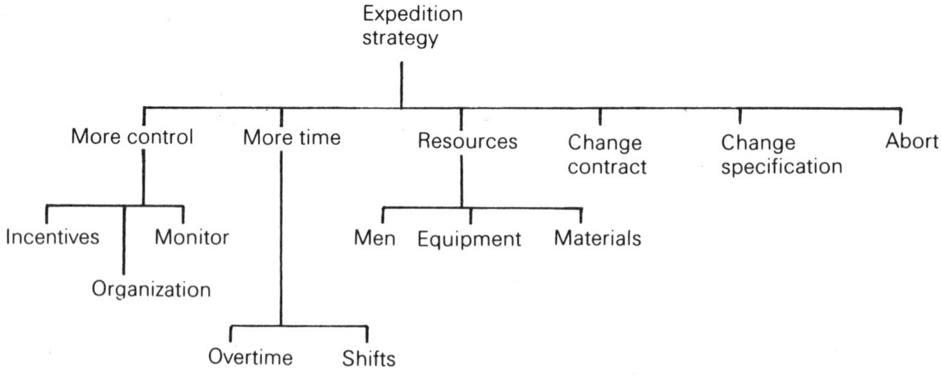

Figure 8.23 Hierarchy of generic expedition strategies.

however, if the next size up were available it could be used instead. Alternatively the framework could be redesigned to accept the next smaller size.

Abort is the final and usually the most drastic measure. If an activity is impossible to achieve, it might be possible to remove it completely from the jurisdiction of the project, treating it as a completely separate scheme. An example occurs in projects in which the objective is to make a bid proposal for a large project. One activity might be to obtain subcontractor quotes which are used to estimate costs. However, if pushed for time, the estimation can be based on the cost of previous work, with appropriate factors to account for the amount of work and inflation.

This hierarchy of strategies can be modeled using the associative network scheme as a type-hierarchy. The six generic strategies are all types of expedition-strategy and below them are subtypes. Below the subtypes are the specific instances of expedition strategies, e.g. 'overtime1' which is a specific instance of overtime work and will typically have attributes of 'number_of_hours', and so on. The generic expedition model fits into the generic project model as shown in Figure 8.24; also shown is an example of how a specific activity may be expedited.

Clearly the suitability of any of these strategies to particular activities requires an insight into the nature of the activities, and the state of the project. This is what the project model is able to describe (section 8.5). The knowledge of strategy suitability can be coded as rules which are gleaned from the expert; an example might be:

```
IF      the task is labor-intensive
AND     the company allows shift schemes
AND     the duration of the task is not short
AND     there is an excess of that labor type
AND     it is easy to hand over work from shift to shift
THEN
        suggest shift work as a strategy
```

Rules like this, for all strategies, have to be obtained by extensive discussions with the expert project manager. This is not an easy task, and it is not feasible to have to repeat it for every different project. Therefore, every effort is made to design the rules to be as generic as possible by finding fundamental attributes which apply to the generic class of projects the system will encounter.

The consultation session is then characterized by reasonable questions which are largely project-independent, like 'on this activity, is it easy to hand over work from shift to shift?' Of course this question itself would be better dealt with by the knowledge-based system if it had a knowledge

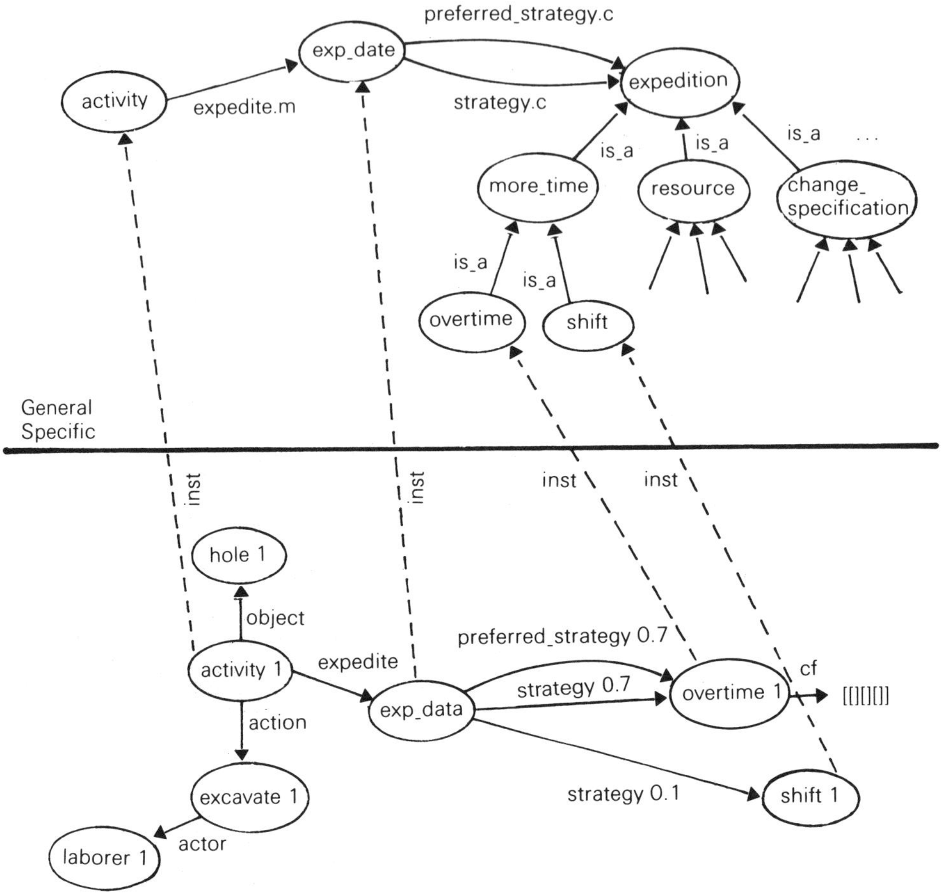

Figure 8.24 Specific and general plans for expedition strategies.

of the kind of aspects which make it difficult to hand over between shifts; this depends on the depth of the analysis adopted by the knowledge engineer.

In their abstract form these strategies can be applied to any project, but some are easier to apply than others. For example, the rules for applying strategies of 'more-time' and 'manpower-resources' are similar for a wide range of projects; fortunately, these tend to be the most popular strategies. Strategies like 'change-specification' are very dependent on the class of project; therefore XPERT requires detailed knowledge of how these types of strategies can be applied. For XPERT to be effective it has to be targeted at a range of projects – probably quite narrow (e.g. proposal projects), where general rules on popular strategies can be refined, and particular rules regarding special expedition strategies incorporated.

One fundamental assumption implicit in the taxonomy of expedition strategies is that the topology of the project network remains unchanged: in other words, replanning is not a valid strategy. In practice, however, replanning is a valid strategy, and one which is applied. Replanning in the real world as a means of expedition is about changing the logical dependency between activities. This is contingent upon the nature of the dependency; in some cases the dependency can be relaxed. Some examples of replanning follow.

It is possible that the original plan simplified the true nature of a logical dependency. For example, suppose that normally the construction of an assembly would not proceed until all the components are available. However, in practice, if certain parts are missing, they can be readily appended later. The strategy is to build what is possible now and create an extra activity to complete the assembly when the missing parts arrive. Typically, the CPA network would not detail the logical dependency of all components, since the high-level plan is usually sufficient; but in a crisis, the task can be examined in detail and replanned. Often there is an additional risk associated with replanning; for example, building on a foundation that has not been cleared by the building inspectors. In this case the inspection activity is taken out of the sequence, and allowed to happen in parallel with the building task. However, if the inspection fails, the building work might have to be scrapped, causing considerable delay.

The issue of replanning was raised with the project managers at Babcock Contractors. Their approach had always been to generate conceptual plans and avoid absolute detail; that exact method for implementing an activity need not be stated, hence the network rarely needs replanning. Replanning at a detailed level can only be done with a deep knowledge of activities. It is reasonable to suppose that with the expert system, it should be possible to incorporate replanning expertise. However, replanning is not included for the following reasons:

1. It is not desirable to change the project plans, if this is possible, because they represent the preferable sequence of events for good reason. Changing the plans may incur additional risk, and could backfire with costly consequences. Furthermore, there is usually a large overhead in revising the plans; it can take some time before the updated plans accurately reflect the state of the project, and the value of an inaccurate plan is questionable.
2. In the field of artificial intelligence, problems associated with planning are known to be extensive, to the extent that planning is recognized as a separate field within AI. In observing the time-scale, it was considered over-ambitious to include this issue.

3. Replanning as an expedition strategy is a different problem from that of expediting a project without changing its topology.

   The algorithms developed from operations research address the traditional expedition problem; possibly some of the AI planning techniques can address the replanning problem. In targeting the XPERT solution to the first problem no more assumptions are made than other researchers in this field.
4. Replanning is a global strategy, not one which applies to individual activities, and is not therefore consistent with the strategies in the expedition hierarchy.
5. The PIPPA project addressed the problem of project planning (and project replanning). Therefore the work with XPERT was of a complementary nature and every effort was made to ensure that the planning system and the expedition system were compatible, with the long term aim that the two tools should be integrated.

Having derived an expedition strategy for an activity, the next step is to estimate the time–cost trade-off function. For some of the expedition strategies this estimation is quite automatic. Consider the strategy of overtime: information regarding the number of men at work, their rates of pay, the overtime rate, duration of the activity and so on, is used to estimate the cost function based on costing equations. Fortunately it is the strategies which occur most frequently which tend to be automatically estimated in this way, namely: overtime, shift work, additional resources. For those strategies which cannot be automatically estimated, like 'change-contract', the system prompts the user for the different delivery dates and associated costs. Figure 8.25 shows a schematic of the XPERT system and Figure 8.26 shows a typical screen display following a consultation with the system.

Automatic estimation is a major asset of the system since it relieves the project manager of the tedium he would otherwise face in trying to calculate these results himself. Once the costing equations have been validated, they effect a reliable estimating mechanism, which accesses data from the information base (if RBFS were linked to the company databases the process would be totally transparent). The reader is referred to Barber and Boardman (1988a) to see how the major expedition strategy cost equations are derived. These are parametric equations of time and cost with some variable dependent on the strategy, e.g. number of hours worked overtime, number of extra men. Due to the discrete nature of the variables, the resultant time–cost trade-off functions are also discrete.

The technique of project expedition is not only valuable when applied

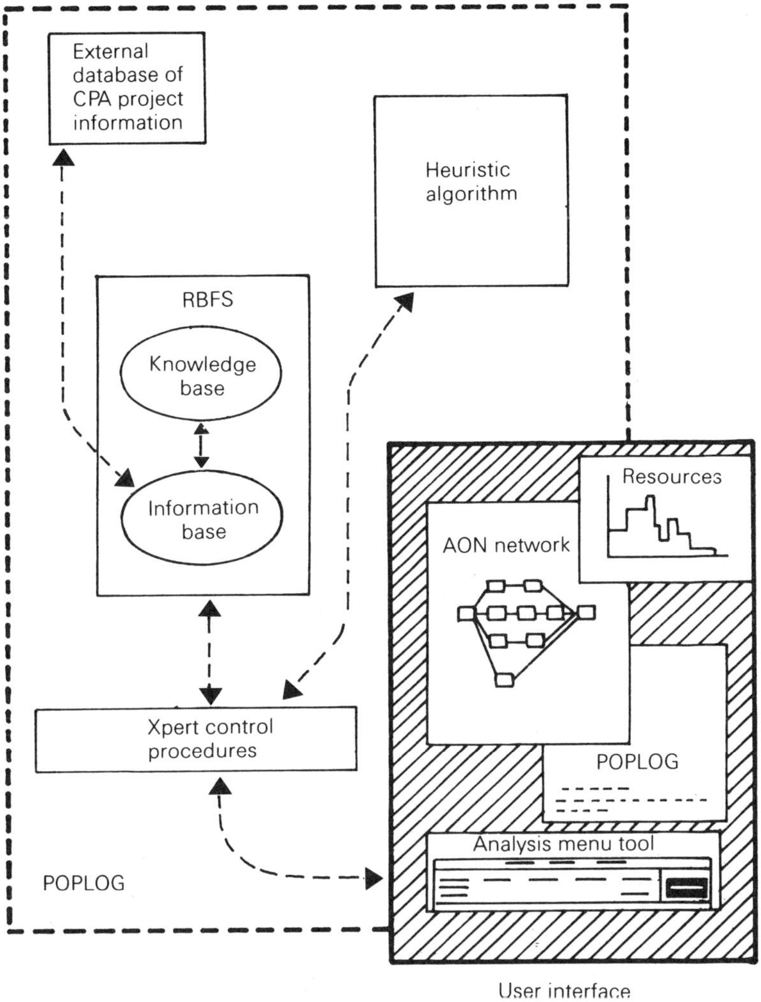

Figure 8.25 Schematic of Xpert system.

to wayward projects. Indeed, there is virtue in the application of the technique to facilitate the optimization of project plans, at the planning stage. The project completion date as yielded by the CPA calculations is not necessarily the optimal completion date in terms of total project cost. Expedition can be employed to reduce the project duration, but characteristically this incurs an increase in direct costs. However, a reduction in the project duration leads to a saving in indirect costs, e.g. site overheads. Furthermore, suppose that the project is to build a process plant

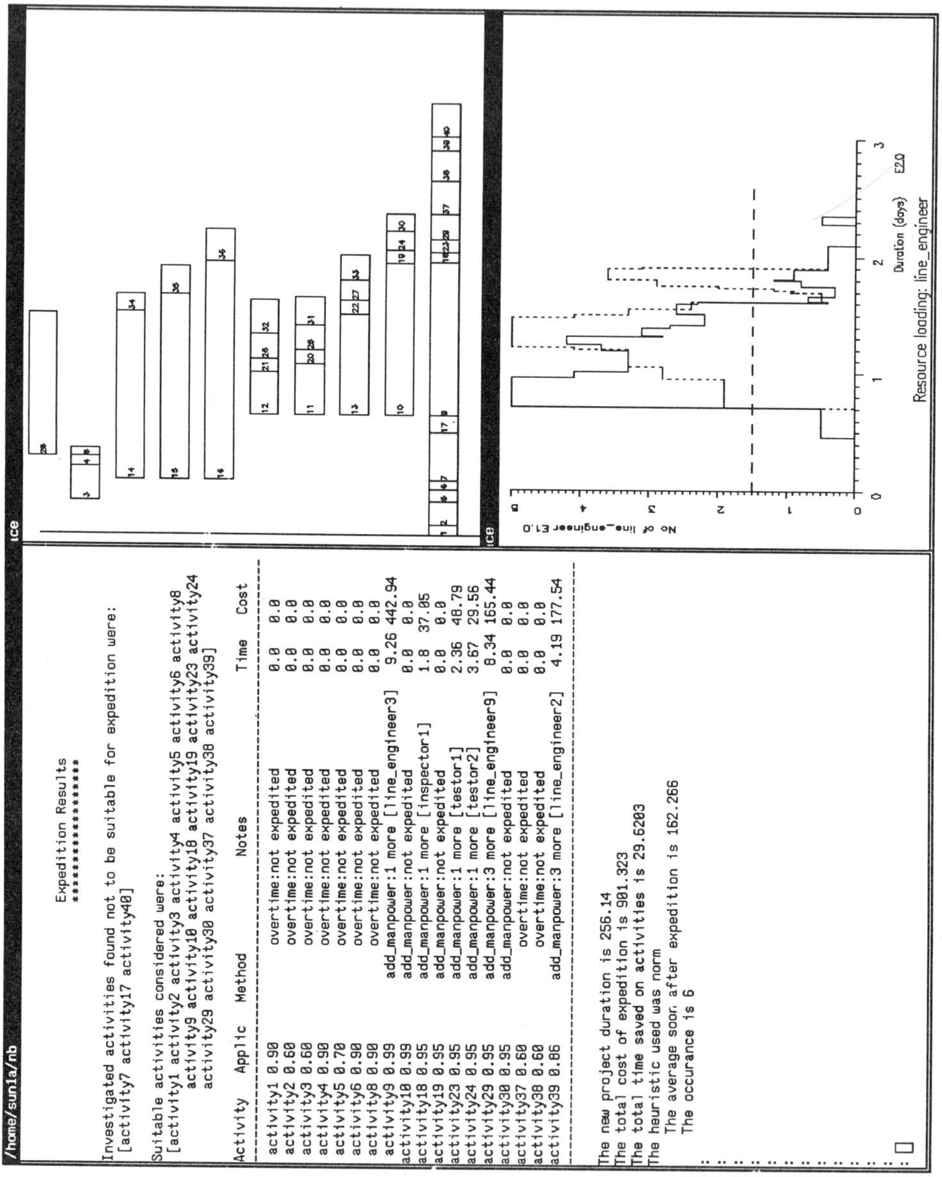

Figure 8.26 The XPERT screen after the execution of the algorithm.

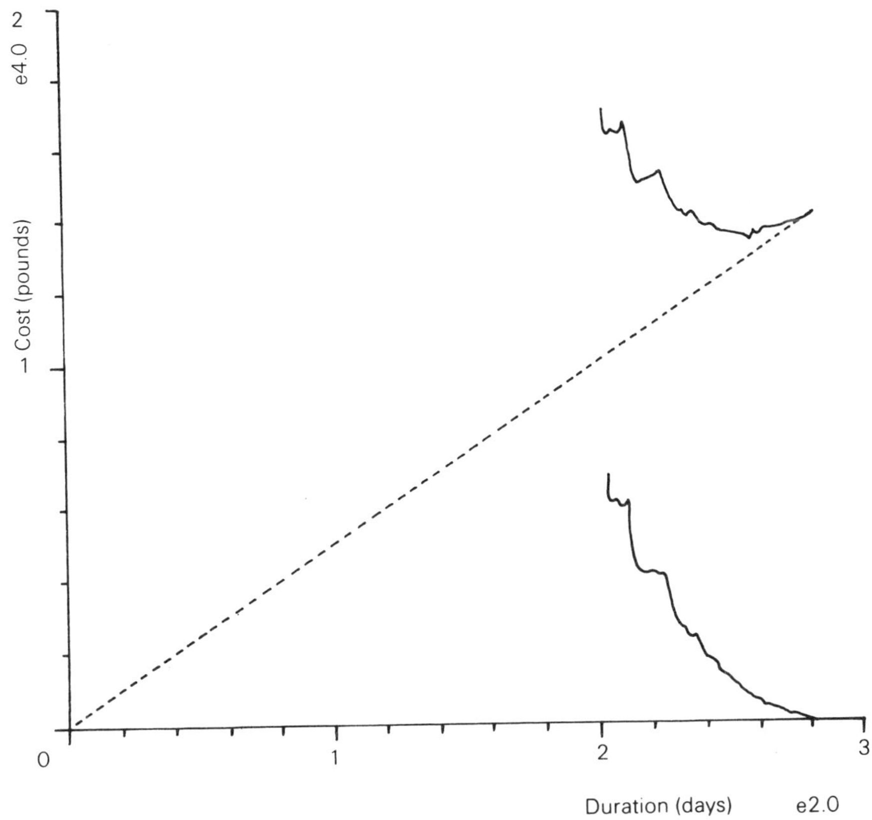

Figure 8.27 Total project cost as a function of project duration.

which will earn revenue when operational, then expedition would bring the plant on-line sooner, thereby reducing interest payments, expediting revenues and so leading to earlier profitability. It is therefore quite possible that project expedition could be used to determine a project with a lower total cost than the original plan. This evokes the notion of an optimal project duration, which has the lowest total cost. To determine the optimal duration requires the calculation of two functions: firstly, the saving in indirect costs (including any additional early revenue); and secondly the cost of expediting the project, both a function of project duration. The optimal project duration is then found from the minimum of the sum of the two functions; Figure 8.27. In this case, site overheads are assumed to be £50 per day, and it can be seen that the minimum cost point, and hence the optimal project duration, is around 260 days.

8.7 REPRISE

The closing remarks of this book are centered upon the philosophical treatment of systems *per se*, an intellectual framework which is of great benefit to engineers in the design of both physical and abstract systems, and something which provided an invaluable underpinning for the AI planning research cited above.

Systemicity is a measure of the degree to which an entity fulfills the description of a system. For an entity to be regarded as a system, there is general agreement that it should exhibit the following three attributes:

1. Be distinct.
2. Be meaningful *per se*.
3. Display cohesion.

It is suggested therefore that the essential components of systemicity should be systemic concepts which convey these attributes, and three such fundamental concepts are proposed, those of boundary, emergence and hierarchy.

The first of these, the *boundary*, when constructed, differentiates the system from its environment, or as some prefer, the wider system and then the environment. Systems designers, creators or synthesizers of systems must decide where the boundary lies – a non-trivial exercise for systems and software engineers, among others. Systems analysts, of whom scientists are a prime example, must discover where the boundary 'actually' lies, when it is not immediately 'obvious'. In investigative mode, scientists create experiments and decide, as designers, where the boundary lies, by definition. Sometimes they show great surprise when the results of their experiment do not satisfactorily transfer to the real world, and think only as a last resort of the location of their boundary. Accident investigators have the incredibly difficult task of deciding where the boundary lies in order to make sense of a tragedy which may prove to have been needless. The concept 'boundary' is so fundamental that from practical considerations it affects both the meaning and cohesion of an entity. A boundary constructor, a 'device' for the location of a system's boundary, is by far the most elusive of intellectual constructs, and the most powerful in the hands of those who can grasp it firmly.

Emergence, the second systems concept, is that which makes sense of describing an entity as having an existence and a meaning in its own right as opposed to calling it a collection of other things. Is a tree meaningful or is it a collection of branches, twigs, leaves, roots and the odd trunk? Well now. What do birds make their nests in? What do poets sit under for inspiration? What do children climb as they scream for their Jane (or Tarzan)? What do lumberjacks fell as they sing merrily in comedy sketches? To each of these subjects the tree is exceedingly meaningful, not simply a collection of branches, twigs and the odd trunk.

There comes a point (in time) or a level (in hierarchy) when it must be acknowledged or recognized that something new has come into existence (and come to stay). That being so, think on this: what emergent feature does a forest possess, or is there no such thing but simply a collection of trees(?), glens and the odd undergrowth? This exercise adumbrates the essential difficulty of defining the emergent features of a system. Pity the system designer whose responsibility is so to construct the boundary that the resultant system achieves all that is desired and yet exhibits nothing unforeseen that leads to failure. There are those who argue that failure is an output built into every system and that this output is set inevitably when the appropriate input conditions obtain!

The third and final concept is *hierarchy* and this has already been alluded to in the preceding paragraph. Accepting emergence as a fundamental concept, we can go on to say that the leaf is part of the tree, the tree is part of the forest, the forest is part of the countryside and so on. But we must ask, what is the nature of being a part, how do the parts interrelate, what do the interrelationships do to contribute to the continued existence of the system and what does the system do to maintain the parts and their interrelationships?

Hierarchy conveys the very essence of structure in a system's constitution; it suggests an interaction of parts, controlled in some way, that sustains a system and thereby enables the system to take care of its parts; it implies a form of government either natural or designed whereby parts, interdependencies and system cohere and thereby provide emergence and portray boundary. Systems designers marvel at the governmental laws and principles exhibited in nature, and strive to impersonate them in their artefacts, be these roads or cities, satellites or telecommunication networks, forges or factories, subroutines or operating systems.

The fact that these concepts are simple to grasp, appearing almost self-evident, but prove difficult to put into practice, should not be a disincentive. There can be no other way to design than to do so systemically. That is a lesson from nature. Systems analysts should strive more fervently to understand these concepts in ways that will help them find methodologies which are easy for their users to understand and by means of which these concepts can be put to practical use, in order that entities can be produced which look less like catastrophes and more greatly resemble systems. Perhaps that is what *systems* engineering is all about. If so, then its advocates ought to be prepared to defend their respective methodologies against the measure of systemicity. I have used these systemic concepts to good effect in the major program of research and development that I have been describing. In this it was essential to combine simplicity and complexity at one and the same time. The complexity has come from the problem being addressed, namely the application of artificial intelligence (knowledge-based systems techniques) to project management. The simplicity had to be added to the program by the R&D personnel so that their collaborators could understand it – R&D credibility was always an endangered feature, put further at risk by the considerable hype which surroundeds AI, especially its application to practical issues.

The foregoing remarks have typified the thinking of what has been termed the 'systems movement', which of course is but one example of the many 'movements' in which mankind has been involved. Perhaps the primitive form of human movement is epitomized by man venturing from his cave in the first instance. Ever since that preliminary excursion, man has continued to adventure, 'moving' into sophisticated forms of transportation, communication and shelter. It is this development that Bronowski describes as 'the ascent of man'. In contemporary society, exemplars of movements are found in symphonies, cults, meteorology and holidays; a veritable eclectic collection, typifying man's striving to fulfill his desires.

The systems engineer has contributed to the movement of mankind in its upward climb to a better life. He continues to be inventive in his design of modern tools, which are benefiting in their conception and design from the sophisticated theories of the systems movement. At the same time, these tools, in the hands of the skilled systems engineer, are shaping the future of the world in which we are to live. After all, systems engineering is an immensely practical subject.

8.8 EXERCISES

1. 'Artificial intelligence is an experiment to aid us in the discovery of the truth about ourselves. Systems engineering is the means by which we come to terms with the truth in the real world.' Discuss.

2. Choose an area of interest to you, and then describe how you, as a knowledge engineer, would go about building an expert system for that domain. You

should include in your description references to the kinds of tool you need, the modeling of the knowledge to be elicited, the means for acquiring that knowledge, and the process of verifying and validating the resultant expert system. What do you suppose this exercise would provide by way of new thinking for the community of expert system builders? How do you think this expert system could be marketed effectively.

3. Using the methodology described in section 8.3, which described the philosophy and architecture of the rule-based frame system, conceptualize one of the following:

(a) a house buying exercise;
(b) a choice of routing between two locations;
(c) making dinner for the boss and his wife.

Your model should indicate the fundamental concepts and their interrelationships, and the knowledge required in order to achieve a specified mission statement.

4. Imagine a tableau which sets out the interrelationships between the functions and the levels of project management. The former consist of objective-setting, plan generation, resource scheduling and project control. The levels are defined as executive (upper), work-package (middle) and task (lower). Choose a particular project management scenario from one of the following (or other of your choice as you prefer), and then describe the kinds of artificial intelligence tools and techniques would best support the elements of this tableau, stating the reasons for your choice:

(a) construction of the Channel tunnel;
(b) launch of the NASA space station;
(c) shutdown planning for a methanol plant.

5. Using the methodology described in section 8.5, develop a project planning methodology, to whatever depth you choose, for any of the scenarios offered in exercise 4.

6. 'Corporations of the future will be advanced users of all forms of information technology, including office automation, desk-top publishing, client databases, intra-personnel and human-equipment communication systems. Planning will become the vital mission – a distributed yet integrated function, which unites the entire workforce. Expert planning systems will therefore provide the backbone of corporate planning.' Discuss.

8.9 FURTHER READING

Barber, T.J. (1988) 'Project control strategies using a heuristic algorithm and intelligent knowledge-based system', Ph.D. Thesis, CNAA.

Barber, T.J. and Boardman, J.T. (1988a) 'Heuristic optimisation techniques in knowledge based project control', Proceedings of 12th IMACS Conference, Paris, July.

Barber, T.J. and Boardman, J.T. (1988b) 'Knowledge-based project control employing heuristic optimisation', *IEE Proceedings*, **135A** (8), 529–38.

Barber, T.J., Marshall, G. and Boardman, J.T. (1988) 'A philosophy and architecture for a rule-based frame system (RBFS)', *International Journal of Engineering Applications of Artificial Intelligence*, **1** (June), 67–85.

Boardman, J.T. (1988) 'Al in engineering planning and control', Symposium on Information Technology for Nuclear Plant, Gloucester Hotel, London, 5–6 October.

Boardman, J.T. (1989a) 'Applications of artificial intelligence to the building services industry', Invited paper to CIBSE, held 19 September.

Boardman, J.T. (1989b) 'Project planning and control: the role of expert systems', Proc. IEE vacation school on the applications of artificial intelligence, University of Reading, September.

Boardman, J.T., Brown, N. and Kuczora, P.W. (1988a) 'Engineering project management using expert planning systems', Invited paper presented at Wright Patterson Air Force Base, Dayton, Ohio, 19 October.

Boardman, J.T., Brown, N., Kellett, J.M. and Kuczora, P.W. (1988b) 'Expert planning systems in engineering management', Proceedings of the Symposium on Management of High-Tech Projects, Hyatt Regency Hotel, Singapore, 25 November.

Bobrow, D.G. (1975) 'Dimensions of representation', in Bobrow, D.G. and Collins, A. (Eds.) *Representation and Understanding*, pp. 2, 34, Academic Press.

Buchanan, B.G. and Shortliffe, E.M. (1984) *Rule Based Expert Systems*, p. 132, Addison-Wesley.

Charniak, E. and McDermott, D. (1985) *Introduction to Artificial Intelligence*, Addison-Wesley.

Feigenbaum, E., McCorduck, P. and Penny Nii, H. (1988) *The Rise of the Expert Company*, Times Books.

Findler, N.V. (1979) *Associative Networks – Representation and Use of Knowledge by Computer*, Academic Press.

Garrett, C. (1989) 'Functional diagnostic strategies for analogue systems using heuristic programming techniques', Ph.D. Thesis, CNAA.

Goldstein, I.P. and Roberts, R.B. (1984) 'NUDGE – a knowledge-based scheduling program', in *Artificial Intelligence – an MIT Perspective*, Vol. 1, pp. 255–84, MIT Press.

Hayes-Roth, B. (1985) 'Blackboard architecture for control', *Journal of Artificial Intelligence*, **26**, 251–321.

Kellett, J.M. (1986) *VEGAN: Visual Editor for the Generation of Associative Networks: a knowledge engineering tool*, ITRI research report No. 16.

Kellett, J.M. and Boardman, J.T. (1989) 'Knowledge-based system for flight simulator configuration, project planning and resource allocation', Proceedings of the 2nd Colloquium on the application of IKBS for Computer Integrated Manufacturing, Boulevard Hotel, Singapore, 21 July.

Lea, S.M. (1988) 'Expert system for the functional test program generation of digital electronic circuit boards', Proc. International Test Conference, Washington, DC, September.

Marshall, G. (1988) 'Synthesis of project plans using an intelligent knowledge-based systems methodology', Ph.D. Thesis, CNAA.

Marshall, G., Barber, T.J. and Boardman, J.T. (1987a) 'A methodology for modelling a project management control environment', *IEE Proc.*, **134A** (4).

Marshall, G., Kellett, J.M. and Lim, B.S. (1987b) 'Planning systems in manufacturing engineering', Proceedings of the KBS 87 Conference, London, 23–25 June.

Minsky, M. (1975) *A Framework for Representing Knowledge, The Psychology of Computer Vision*, McGraw-Hill.

Penny Nii, H. (1986) 'Blackboard systems', *AI Magazine*, part I in **7**, (2), 38–53, and part II in **7**, (3), 82–106.

Raj Reddy, D., Erman, L.D. and Neely, R.B. (1973) 'The HEARSAY speech understanding system: an example of the recognition process', *IJCAI* **3**, 185–93.

Waterman, D.A. (1986) *A Guide to Expert Systems*, Addison-Wesley.

Winograd, T. (1975) 'Frame representations and the declarative/procedural controversy', in Bobrow, D.G. and Collins, A. (Eds.) *Representation and Understanding*, pp. 45–78, Academic Press.

Winston, P.H. (1977) *Artificial Intelligence*, Addison-Wesley.

Winston, P.H. and Horn, B. (1980) *LISP*, Addison-Wesley.

Index